INTERNATIONAL CENTRE FOR MECHANICAL SCIENCES

COURSES AND LECTURES - No. 27

RASTKO STOJANOVIĆ

UNIVERSITY OF BELGRADE

RECENT DEVELOPMENTS IN THE THEORY OF POLAR CONTINUA

COURSE HELD AT THE DEPARTMENT
FOR MECHANICS OF DEFORMABLE BODIES
JUNE - JULY 1970

UDINE 1970

SPRINGER-VERLAG WIEN GMBH

ISBN 978-3-211-81144-3 ISBN 978-3-7091-4309-4 (eBook)
DOI 10.1007/978-3-7091-4309-4

Originally published by Springer Vienna in 1970

Preface

This course of 21 lectures on RECENT DEVELOPMENTS IN THE THEORY OF POLAR CONTINUA is based mostly on the former course I had the pleasure to give at the International Centre for Mechanical Sciences in Udine during the September-October session in 1969. Being aware of many important topics in the mechanics of polar continua which I did not include in my former course, and even more being aware of the plenty of mistakes, and most of them were not of the typographical nature, I was very glad to receive the invitation of the Rector of CISM to give another course of lectures on the same subject as I did nine months ago.

Owing to the lack of time at home, I had to prepare this course in Udine. The Rector, W. Olszak, the Secretary General, Professor L. Sobrero, together with the complete technical and administrative staff of CISM did everything possible to make my stay and work here not only efficient, but also a pleasure. I mostly admire their support and assistance.

In parallel to this course, at CISM were held the courses by the most distinguished scientists, Professors Eringen, Nowacki, Mindlin and Sokolowski on more specialized topics of polar continua. Therefore, I have omitted from my lectures the chapters dealing with the applications of linearized theories to some special problems, such as wave propaga-

tion, stress concentration, singular forces etc. I have added a chapter on some aspects of the shell theory, and 2 chapters on polar fluids and on the theory of plasticity, as well as some other minor corrections and additions (e.g. on incompatible strains with applications to thermoelasticity and to the theory of dislocations). Also the list of references is corrected and the references are also given to some recently published papers.

I mostly appreciate the help of Mr.J. Jarić, M. Sc. in correcting the list of references and in checking the proofs in the main text, as well as the help of Mr. M. Micunović, B. Eng. in writing the formulae.

The International Centre for Mechanical Sciences in Udine paid for the second time in one year its attention to the mechanics of polar continua. Appreciating very much this interest in this modern branch of mechanics, I hope that this course of lectures (which I delivered with the greatest pleasure) will be, besides all imperfections and may be even conceptual errors, at least a small contribution to the further development of continuum mechanics.

Udine, July 16, 1970

R. Stojanović

1. Introduction

Classical continuum mechanics considers material continua as point-continua with points having three degrees of freedom, and the response of a material to the displacements of its points is characterized by a symmetric stress tensor. Such a model is insufficient for the description of certain physical phenomena.

Already in 1843 St. Venant [471] * remarked that for the description of deformations of thin bodies a proper theory cannot be restricted to the analysis of deformations of a straight line which can be only lengthened and bent, but must also include directions which can be rotated independently of the displacements of the points.

A further generalization of this idea was to attach to each point of a three-dimensional continuum a number of directions which can be rotated independently of the displacements of the points to which they are attached. That physical bodies might be presented in this way was suggested in 1893 by Duhem [94]. In the study of crystal elasticity Voigt [473, 474] came to the same ideas. It is the merit of the brothers Eugène

---

* The numbers in square brackets refer to the list of references at the end of these lectures.

and François Cosserat that a theory of such oriented continua was developed, and there are three papers by them[71, 72, 73] published in 1907-1909 which are the basis of all later work on polar continua. However, their work remained forgotten until 1935, when Sudria [437] gave a more modern interpretation of their theory, applying the contemporary vectorial notation.

One of the essential features of polar continua is that the stress tensor is not symmetric, and the well known second law of Cauchy is to be replaced by another one from which the Cosserat equations follow.

In oriented bodies the antisymmetric part of the stress tensor, according to the Cosserat equations, is related to the divergence of a third-order tensor of couple-stresses. This tensor, through the constitutive relations, depends on the deformations of the directors, but the deformations of directors are not the only deformations responsible for the couple-stresses.

The non-symmetry of the stress tensor appears also if the higher order deformation gradients are taken into account, instead of the first-order gradients only, as it is the case in the classical continuum mechanics. According to Truesdell and Toupin [467], Hellinger [202] was the first in 1914, to obtain the general constitutive relations for stress and couple-stress, generalizing an analysis of E. and F. Cosserat.

In 1953 Bodaszewski [39] developed a theory of non-

-symmetric stress states, but without any reference to earlier works. He applied the theory to elasticity and fluid dynamics.

Since 1958, the general interest in the non-symmetric stress tensor and in the Cosserat continuum rapidly increases. In that year Ericksen and Truesdell published a paper on the exact theory of rods and shells in which they considered a generalized Cosserat continuum, i.e. a medium with deformable directors, but without any constitutive assumptions. Günther [190] gave a linear theory (statics and kinematics) of the Cosserat continuum, with a very interesting application to the continuum theory of dislocations, and Grioli [179] developed a theory of elasticity with the non-symmetric stress tensor. Ericksen's theory of liquid crystals and anisotropic fluids is also an application of the theory of oriented bodies [101].

There are different physical and mathematical models of continua which serve as generalizations of the classical concept of a point continuum. All such models in which the stress tensor is not symmetric are regarded here as POLAR CONTINUA.

2. Physical Background

It was already mentioned that the classical model of a material continuum is insufficient for the description of a number of phenomena. In the case of thin bodies this can already be seen.

If we regard a very thin circular cylinder, a one-dimensional representation is the sufficient approximation for the study of its elongation, but twists are excluded from such considerations. In order to include the twist we may associate a unit vector with each point of the line, and rotations of this vector give us the needed information on the twist. Obviously, this rotation is independent of the displacements of points of the line.

For the study of a flexible string a rigid triad of unit vectors may be attached to each point of the string.

In the theory of rods, plates and shells the situation is similar. In the direct approach to the theory of rods, Green and Laws [153, 156] define a rod as a curve at each point of which there are two assigned directors. The theory of plates and shells may be based on the model, consisting of a deformable surface with a single director attached to each of its points. Such a surface is called by Green and Naghdi [165] a Cosserat surface.

A crystal in the continuum approximation is a point continuum, but the rotations of particles cannot be represented in such an approximation. In order to include the interactions of rotating particles in crystal elasticity, Voigt [473, 474] was the first to generalize the classical concepts of continuum mechanics.

Ericksen [105] developed the theory of liquid crystals and anisotropic fluids assuming that a fluid is an ordinary three-dimensional point continuum with one director at each point. Particles of the fluid are assumed to be of the dumb-bell shape.

Continuum mechanics is a method for the study of mechanical properties of bodies the dimensions of which are very great in comparison with the interatomic distances. The discrete structure of matter, in fact, is to be studied if we wish to make an exact theory of the behaviour of matter. For bodies containing a large number of particles it is practically impossible. The classical point continuum is just an approximation, and some models of continua are constructed in such a way to represent a better approximation and to include some effects which can not be interpreted from the point of view of a point continuum.

In a series of papers Stojanović, Djurić and Vujosevic [428] in 1964, Green and Rivlin (for references see Rivlin [378]) have taken as the starting point the discrete structure of particles which constitute the medium. Each particle consists of a number of mass-points. The continuum representation consists of a point continuum, the points correspond to the centres of gravity of particles, and in a number of deformable vectors, the directors. The distribution of masses in such a representation is specified through some inertia coefficients. The forces acting on mass-points in the continuum representation reduce to the simple forces acting on the points of the continuum and on the director forces acting on the directors, as well as to the simple and director surface forces (stresses) and couples, measured per unit area of the deformed surface.

Kröner, Krumhansl, Kunin and other authors approach this problem of approximation from the point of view of solid state physics [255] . We shall mention here only the very impressive picture of the couple-stress given by Kröner in a dislocated crystal [252] . From the distribution of microscopical stresses, applying an averaging process, Kröner computed the macroscopic moments. The obtained couple-stress he attributed to the non-local forces, i.e. to the long-range cohesive forces.

Mindlin [285] and Eringen and Suhubi [138] introduced microstructure into the theory of elasticity and into continuum mechanics, in general. The unit cell of material with microstructure might be interpreted as a molecule of a polymer, as a crystalite of a polycrystal, or as a grain of an incoherent material. The concept of microstructure Eringen introduced also into the fluid mechanics [121] .

Eringen generalized further the model and defined micromorphic materials [125]. A volume element of such a material consists of microelements which suffer micromotions and microdeformations. Micropolar materials are a subclass, in which the microelements behave as rigid bodies.

The theory of multipolar media by Green and Rivlin [172, 173] represents a very fine abstract and general mathematical treatment of generalized continua, from which many theories follow as special cases.

Besides the physical models mentioned which served as a basis for different continuum-mechanical representations, there is a number of other theories and treatments inspired by the problems of solid-state physics (Teodosiu [449]), or by the structure of technical materials (Misicu [306]) or by the mathematical possibilities for generalization of classical concepts (Grioli [179], Aero and Kuvshinskii [5]).

Granular media represent also the field in which the methods of generalized continuum mechanics are applied (Oshima [347]).

It is impossible to mention all contributors to the contemporary development of continuum mechanics, and we restricted this list only to some of them whose work most inspired further research.

3. Motion and Deformation

We shall regard material points as the fundamental entities of material bodies.

A body **B** is a three-dimensional differentiable manifold, the elements of which are called material points. *

* This definition of a body corresponds to the definition given by Truesdell and Noll [468]. Noll [330] developed a very general approach to continuum mechanics, but we are not going to follow

The material points $M_1, M_2, \ldots$ may be regarded as a set of abstract objects M mentioned in the Appendix, section A1, so that the 1:1 correspondence of the points M_k and of the points of a three-dimensional arithmetic space establishes a general material three-dimensional space. Since bodies are available to us in Euclidean space, we shall relate the points M_k to the points of Euclidean space, establishing a 1:1 correspondence between the points M_k of a body B and points $\underset{\sim}{x}$ of a region R of this space. The numbers x^i, $i = 1,2,3$ represent coordinates of the material point M and the points $\underset{\sim}{x}$ are places in the space occupied by the point M.

Any triple of real numbers $x^i, i = 1,2,3$ may be regarded as an arithmetic point, which belongs to the arithmetic space A_3. A 1:1 smooth correspondence between the material points M of a body B and arithmetic points $\underset{\sim}{X}$, such that $X^K = X^K(M)$, $K = 1,2,3$ represents a system of coordinates in which individual material points are characterized by their <u>material coordinates</u> X^K, $K = 1,2,3$.

A 1:1 correspondence between points $\underset{\sim}{x}$ of a region R of Euclidean space, and points M of a body B is the configuration of the body,

---

it since it does not include plasticity and mostly is concerned with the non-polar materials, regarding elasticity, visco-elasticity and viscosity from a unique point of view. For the general approach to this theory, because of its highest mathematical rigour and for a very complete bibliography we refere the readers to the book by Truesdell and Noll [468].

$$x^i = x^i(M) = x^i(X^1, X^2, X^3) . \tag{3.1}$$

The points x^i represent places in the space occupied by the material points M and we shall refer to the coordinates x^i as to the spatial coordinates. The functions $x^i = x^i(\underset{\sim}{X})$ are assumed to be continuously differentiable.

In general no assumptions are made on the geometric structure of the material manifold and it is not to be confused with one of its configurations. It is advantageous to choose one configuration as the reference configuration and to identify material coordinates with the spatial coordinates in the reference configuration.

Thus, the material points of a body B in the reference configuration are referred to a system of coordinates X^k, which is an admissible system of coordinates in Euclidean space, and in the following we shall refer to X^k as to the material coordinates.

Motion of a body is a one-parameter 1:1 mapping

$$x^k = x^k(X^1, X^2, X^3, t) \equiv x^k(X^k, t) , \tag{3.2}$$

or shortly

$$\underset{\sim}{x} = \underset{\sim}{x}(\underset{\sim}{X}, t) ,$$

of the points M in the reference configuration K_0 on the points $\underset{\sim}{x}$ occupied by the material points at a moment of time t, which determines a configuration $K_t = K(t)$. The parameter t is a real parameter and it represents time. We assume that the functions

$\underset{\sim}{x} = \underset{\sim}{x}(\underset{\sim}{X})$ are continuously differentiable.

We assume that

(3.3) $$\det \frac{\partial x^k}{\partial X^K} = \det x^k{}_{;K} \neq 0 \; ,$$

so that there exists the inverse mapping

$$X^K = X^K(x^1, x^2, x^3, t) \; ,$$

short:

(3.4) $$\underset{\sim}{X} = \underset{\sim}{X}(\underset{\sim}{x}, t) \; .$$

The partial derivatives

(3.5) $$F^k{}_K \equiv \partial x^k / \partial X^K \equiv x^k{}_{;K} \; ,$$
$$F_k{}^K \equiv \partial X^K / \partial x^k \equiv X^K{}_{;k} \; ,$$

are called <u>deformation gradients</u>, and the total covariant derivatives (see Appendix, section A3)

$$x^k{}_{;KL}, \ldots, x^k{}_{;K_1 \ldots K_N} \; ,$$
$$X^K{}_{;kl}, \ldots, X^K{}_{;k_1 \ldots k_N} \; ,$$

represent deformation gradients of order $2, 3, \ldots N$.

Let K_0 and K be two configurations of a body B, K_0 referred to material coordinates X^K, and K referred to spatial coordinates x^k. The systems of reference X^K and x^k are chosen independently of one another. The <u>deformation</u> is a mapping of one configuration on the other,

(3.6a) $$x^l = x^l(\underset{\sim}{X}) \; ,$$

$$X^L = X^L(\underset{\sim}{x}) . \tag{3.6b}$$

If dS^2 and ds^2 are squares of the line elements in the configurations K_0 and K respectively,

$$dS^2 = G_{LM} dX^L dX^M , \tag{3.7}$$
$$ds^2 = g_{\ell m} dx^\ell dx^m ,$$

using the mappings (3.6) we may represent the line element of the reference configuration in terms of the coordinates of the deformed configuration and conversely. From (3.6) we have

$$dX^L = X^L_{;\ell} dx^\ell \qquad dx^\ell = x^\ell_{;L} dX^L \tag{3.8}$$

and

$$dS^2 = c_{\ell m} dx^\ell dx^m , \tag{3.9}$$
$$ds^2 = C_{LM} dX^L dX^M .$$

Here

$$c_{\ell m} \equiv G_{LM} X^L_{;\ell} X^M_{;m} \tag{3.10}$$

is the <u>spatial deformation tensor</u>, and

$$C_{LM} \equiv g_{\ell m} x^\ell_{;L} x^m_{;M} \tag{3.11}$$

is the <u>material deformation tensor.</u>

It is always possible to decompose a non-singular matrix $\mathbf{M}$ into one symmetric and one positive definite matrix,

$$M^{k}_{\cdot\ell} = R^{k}_{\cdot i} S^{t}_{\cdot\ell} = S^{*k}_{\cdot t} R^{t}_{\cdot\ell} \,, \tag{3.12}$$

where $\mathbf{R}$, $\mathbf{S}$ and $\mathbf{S}^*$ are uniquely determined (cf. Ericksen [100], § 43). Applying this <u>polar decomposition theorem</u> to the matrix $\mathbf{F}$ (cf. [468]) of deformation gradients, we obtain

$$\mathbf{F} = \mathbf{R}\cdot\mathbf{U} = \mathbf{V}\cdot\mathbf{R} \tag{3.13}$$

where $\mathbf{R}$ is orthogonal, and $\mathbf{U}$ and $\mathbf{V}$, determined by

$$\mathbf{U}^2 = \mathbf{F}^T\cdot\mathbf{F} \qquad \mathbf{V}^2 = \mathbf{F}\cdot\mathbf{F}^T \tag{3.14}$$

are the <u>right</u> and the <u>left stretch tensors</u>, respectively. The deformation tensors $\mathbf{C}$ and $\mathbf{B}$

$$\mathbf{C} = \mathbf{U}^2 = \mathbf{F}^T\mathbf{F} \tag{3.15}$$

$$\mathbf{B} = \mathbf{V}^2 = \mathbf{F}\mathbf{F}^T$$

are accordingly called the <u>right</u> and the <u>left Cauchy-Green tensors</u>.

Since $\mathbf{F} = \{x^{k}_{;K}\}$, the transposed matrix $\mathbf{F}^T$. is determined by

$$\mathbf{F}^T = \{g_{k\ell}\, x^{\ell}_{;L}\, G^{LM}\}$$

and for the components of the tensors $\mathbf{C}$ and $\mathbf{B}$ we have

$$C^K_L = g_{k\ell}x^k_{;L}x^\ell_{;M}G^{KM} = G^{KM}C_{LM} \qquad (3.16)$$

$$B^k_\ell = G^{KL}x^k_{;K}x^m_{;L}g_{m\ell} = \overset{-1}{c}{}^k_\ell = g_{m\ell}\overset{-1}{c}{}^{km} . \qquad (3.17)$$

The tensor $\overset{-1}{\underset{\sim}{c}}$, with the components

$$\overset{-1}{c}{}^{km} = G^{KM}x^k_{;K}x^m_{;M} \qquad (3.18)$$

is the reciprocal of the spatial deformation tensor $\underset{\sim}{c}$,

$$c_{jk}\overset{-1}{c}{}^{km} = \delta^m_j .$$

If a body suffers only a rigid motion, the distances between its points are preserved, there are no deformations and

$$C_{KL} = G_{KL} \qquad c_{k\ell} = g_{k\ell} . \qquad (3.19)$$

The material and the spatial strain tensors are defined by the following formulae

$$E_{KL} = \frac{1}{2}(C_{KL} - G_{KL}) , \quad e_{k\ell} = \frac{1}{2}(g_{k\ell} - c_{k\ell}) , \qquad (3.20)$$

where we denote, as usually, material tensors and material components by capital letters and capital indices, and spatial tensors and spatial components by small letters and small indices.

Velocity of a material point $\underset{\sim}{X}$ is the vector $\underset{\sim}{v}$

with the components

$$(3.21) \qquad v^i = \dot{x}^i = \left.\frac{\partial x^i(\underset{\sim}{X}, t)}{\partial t}\right|_{\underset{\sim}{X}} = \text{const.}$$

In general, if $\underset{\sim}{T} = \underset{\sim}{T}(\underset{\sim}{x}, \underset{\sim}{X}, t)$ is a time dependent double tensor field (See Appendix, section A1 and A3), the time derivatives with the material coordinates X^K kept fixed are called <u>material derivatives</u> and are denoted by a superposed dot. Sometimes it is useful to place the dot above a superposed bar, which denotes upon which quantity the operation of the material derivation is to be performed. For the tensor field $\underset{\sim}{T}$ we have

$$\dot{T}^{K\dots}_{\dots k\dots} = \frac{\partial T^{K\dots}_{\dots k\dots}}{\partial t} + \left(\frac{\partial T^{\dots}_{\dots}}{\partial x^l} - \left\{{t \atop m l}\right\} T^{K\dots}_{\dots t\dots} - \dots\right)\dot{x}^l =$$

$$(3.22) \qquad = \frac{\partial T^{K\dots}_{\dots k\dots}}{\partial t} + T^{K\dots}_{\dots k\dots,l}\,\dot{x}^l .$$

Acceleration $\underset{\sim}{a}$ is a vector with the components defined by

$$(3.23) \quad a^i = \dot{v}^i = \ddot{x}^i = \frac{dv^i}{dt} + \left\{{i \atop j k}\right\} v^j v^k = \frac{\partial v^i}{\partial t} + v^i_{;j}\dot{x}^j .$$

The rate of change of the arc element may be calculated directly from $(3.7)_2$,

$$(3.24) \qquad \dot{\overline{ds^2}} = g_{ij}\left(\dot{\overline{dx^i}}\, dx^j + dx^i\, \dot{\overline{dx^j}}\right) .$$

Since

$$dx^i = x^i_{;L} dX^L ,$$

and the material coordinates are kept fixed, we have

$$\dot{\overline{dx^i}} = \dot{\overline{x^i_{;L}}} dX^L = \dot{x}^i_{;L} dX^L = v^i_{;L} dX^L = v^i_{;k} dx^k \qquad (3.25)$$

and

$$\dot{\overline{ds^2}} = 2v_{j,k} dx^j dx^k = 2d_{jk} dx^j dx^k \qquad (3.26)$$

where

$$d_{jk} \equiv \frac{1}{2}(v_{j,k} + v_{k,j}) = v_{(j,k)} \qquad (3.27)$$

is the <u>rate of strain tensor</u>.

The gradients of velocity $v_{i,j}$ may be decomposed into the symmetric and the antisymmetric part. The antisymmetric part

$$w_{ij} \equiv v_{[i,j]} = \frac{1}{2}(v_{i,j} - v_{j,i}) \qquad (3.28)$$

represents the <u>vorticity</u> tensor.

The tensors of the rate of strain and of the vorticity are mutually independent, but the gradients of these two tensors are related by a simple relation:

$$w_{ij,k} = \frac{1}{2}(v_{i,jk} - v_{j,ik}) \equiv \frac{1}{2}(v_{k,ij} + v_{i,kj} - v_{j,ki} - v_{k,ji})$$

$$= d_{ki,j} - d_{jk,i} = 2d_{k[i,j]} . \tag{3.29}$$

A motion is a rigid body motion if $ds = dS$, and the conditions for a motion to be a rigid body motion are given by (3.19). In terms of the strain tensors these conditions reduce to $\underset{\sim}{E} = 0$ and $\underset{\sim}{e} = 0$. For a rigid body motion the rate of strain vanishes and the velocity field has to satisfy the obvious equations

$$v_{(i,j)} = 0 . \tag{3.30}$$

The conditions (3.30) are necessary and sufficient for a motion to be a rigid motion. If $dx^i = u^i$ is an elementary displacement of a body, from (3.30) it follows that the necessary and sufficient conditions for displacements to determine a rigid motion are

$$u_{(i,j)} = 0 . \tag{3.31}$$

These equations are called Killing equations. In Euclidean space the equations (3.30) and (3.31) are integrable and the integrals represent components of the velocity field and of the displacement field for rigid motions.

Let $d_1\underset{\sim}{R}$ and $d_2\underset{\sim}{R}$,

$$d_1\underset{\sim}{R} = d_1X^K\underset{\sim}{G}_K , \quad d_2\underset{\sim}{R} = d_2X^K\underset{\sim}{G}_K \tag{3.32}$$

be two infinitesimal vectors in the initial configuration of a body. These two vectors determine a surface element $d\underset{\sim}{S}$,

$$d\underset{\sim}{S} = d_1\underset{\sim}{R} \times d_2\underset{\sim}{R} = (\underset{\sim}{G}_K \times \underset{\sim}{G}_L) d_1 X^K d_2 X^L \tag{3.33}$$

with the components

$$dS_M = \underset{\sim}{G}_M \cdot (\underset{\sim}{G}_K \times \underset{\sim}{G}_L) d_1 X^K d_2 X^L \tag{3.34}$$

or, according to Appendix (A1.29),

$$dS_M = \varepsilon_{MKL} d_1 X^K d_2 X^L. \tag{3.35}$$

The surface element may also be represented by an antisymmetric tensor,

$$dS^{PQ} = \varepsilon^{PQM} dS_M = 2 d_1 X^{[P} d_2 X^{Q]}. \tag{3.36}$$

For a surface given by the equations $X^K = X^K(u^1, u^2)$ may choose the vectors $d_1\underset{\sim}{R}$ and $d_2\underset{\sim}{R}$ to have the components

$$d_1 X^K = \frac{\partial X^K}{\partial u^1} du^1 , \quad d_2 X^L = \frac{\partial X^L}{\partial u^2} du^2 \tag{3.37}$$

and from (3.35) and (3.36) we obtain

$$dS_M = \varepsilon_{MKL} \frac{\partial X^K}{\partial u^1} \frac{\partial X^L}{\partial u^2} du^1 du^2 \tag{3.38}$$

$$dS^{PQ} = \frac{\partial X^{[P}}{\partial u^1} \frac{\partial X^{Q]}}{\partial u^2} du^1 du^2 . \tag{3.39}$$

When the body suffers a deformation (3.6), we have

$$\frac{\partial X^K}{\partial u^\alpha} = X^K_{;k}\frac{\partial x^k}{\partial u^\alpha}, \quad (\alpha = 1,2) \tag{3.40}$$

where the equations of the deformed surface are $x^k = x^k[X(u^1,u^2)]$. Introducing (3.40) into $(3.38)_1$ we obtain

$$dS_M = \varepsilon_{MKL}X^K_{;k}X^L_{;\ell}\frac{\partial x^k}{\partial u^1}\frac{\partial x^\ell}{\partial u^2}dx^1du^2 .$$

However,

$$\varepsilon_{MKL}X^K_{;k}X^L_{;\ell} \equiv \varepsilon_{NKL}X^N_{;n}x^n_{;M}X^K_{;k}X^L_{;\ell} = \sqrt{G}\det(X^T_{;t})e_{k\ell n}x^n_{;M},$$

and

$$dS_M = x^n_{;M}\sqrt{\frac{G}{g}}\det(X^T_{;t})\varepsilon_{k\ell n}\frac{\partial x^k}{\partial u^1}\frac{\partial x^\ell}{\partial u^2}du^1du^2 = J^{-1}x^n_{;M}ds_n, \tag{3.41}$$

where

$$J = \sqrt{\frac{g}{G}}\det(x^t_{;T}) . \tag{3.42}$$

Hence,

$$ds_n = \varepsilon_{k\ell n}\frac{\partial x^k}{\partial u^1}\frac{\partial x^\ell}{\partial u^2}du^1du^2,$$

which represents the surface element of the deformed surface. Also

$$ds^{pq} = \frac{\partial x^{[p}}{\partial u^1}\frac{\partial x^{q]}}{\partial u^2}du^1du^2, \tag{3.43}$$

and it may be easily verified that

$$dS^{PQ} = X^P_{;p} X^Q_{;q} ds^{pq} . \tag{3.44}$$

The volume element dv in the initial configuration of a body may be defined in terms of three infinitesimal displacement vectors $d_\alpha \underset{\sim}{R} = d_\alpha X^K \underset{\sim}{G}_K$, $\alpha = 1, 2, 3$,

$$dV = d_1\underset{\sim}{R} \cdot (d_2\underset{\sim}{R} \times d_3\underset{\sim}{R}) = \varepsilon_{KLM} d_1X^K d_2X^L d_3X^M . \tag{3.45}$$

After a deformation we have

$$dV = \varepsilon_{KLM} X^K_{;k} X^L_{;\ell} X^M_{;m} d_1x^k d_2x^\ell d_3x^m .$$

Since

$$\varepsilon_{KLM} X^K_{;k} X^L_{;\ell} X^M_{;m} = \sqrt{G} \det(X^T_{;t}) e_{k\ell m} = \sqrt{\frac{G}{g}} \det(X^T_{;t}) \varepsilon_{k\ell m} ,$$

with the notation $(3.41)_2$ we may write

$$dV = J^{-1} \varepsilon_{k\ell m} d_1x^k d_2x^\ell d_3x^m = J^{-1} dv , \tag{3.46}$$

where dv is the volume element in the deformed configuration,

$$dv = \varepsilon_{k\ell m} d_1x^k d_2x^\ell d_3x^m . \tag{3.47}$$

Some authors, mostly British, prefer the use of <u>convected coordinates</u>, with respect to which the numerical values of coordinates of material points in a deformable body remain unchanged during the motion of the body. Let $X^K = \text{const.}$ be three independent families of material surfaces. At any moment of time these surfaces define a <u>convected system of coordinates</u> for a given motion, and during the motion we have $x^k = X^K\delta^k_K$. To avoid ambiguities we shall denote convected coordinates by θ^k.

If $T_K^{\cdot L}(\underset{\sim}{X})$ is a tensor field in the initial configuration of a body, at time t its components will be

$$T_k^{\cdot \ell}(\underset{\sim}{\theta}) = \delta^K_k \delta^\ell_L T_K^{\cdot L}(\underset{\sim}{X}, t) . \tag{3.48}$$

Since it is no more necessary to distinguish between material and spatial coordinates, it is possible to consider simply the tensor field $T_k^{\cdot \ell}(\underset{\sim}{X}, t)$ which coincides with $T_K^{\cdot L}$ at the initial moment t_0 of time. Thus, the fundamental metric form at time t will be

$$ds^2 = g_{k\ell}(\underset{\sim}{\theta}, t)d\theta^k d\theta^\ell = g_{KL}(\underset{\sim}{X}, t)dX^K dX^L , \tag{3.49}$$

and $g_{k\ell}$ coincides at the initial moment t_0 with the components G_{KL}, and with the components C_{KL} at time t. The strain tensor, with respect to convected coordinates, is defined by

$$e_{k\ell} = \frac{1}{2}[g_{k\ell}(\underset{\sim}{\theta}, t) - g_{KL}\delta^K_k \delta^L_\ell] . \tag{3.50}$$

From (3.49) we have

$$\dot{\overline{ds^2}} = \dot{g}_{k\ell}(\underset{\sim}{\theta}, t)d\theta^k d\theta^\ell , \tag{3.51}$$

and for the rate of strain tensor follows the expression

$$d_{k\ell} = \frac{1}{2}\dot{g}_{k\ell} = \dot{e}_{k\ell} .$$

4. Compatibility Conditions

For a given tensor field $\underset{\sim}{c}(\underset{\sim}{x})$, or $\underset{\sim}{C}(\underset{\sim}{X})$, the deformations

$$\underset{\sim}{x} = \underset{\sim}{x}(\underset{\sim}{X}) \quad \text{or} \quad \underset{\sim}{X} = \underset{\sim}{X}(\underset{\sim}{x}) \tag{4.1}$$

do not necessarily exist. The existence of the deformations depends on the integrability conditions of the equations (3.10) or (3.11), and these conditions are usually called in continuum mechanics the <u>compatibility conditions</u>.

There are six independent equations (3.10), with nine independent deformation gradients $X^K_{;k}$. In order to find the deformations we have first to find the deformation gradients, but since the number of the unknowns, regarding the equations (3.10) as a system of algebraic equations, exceeds the number of equations, we shall first differentiate partially the equations (3.10) with respect to the spatial coordinates x^ℓ, assuming that the deformations (4.1) exist. Thus we obtain a system of 18 equations with 18 unknowns $\partial_m \partial_n X^K$,

$$\partial_\ell C_{mn} = \partial_L G_{MN} X^L_{;\ell} X^M_{;m} X^N_{;n} +$$

$$(4.2) \qquad\qquad + G_{MN}(\partial_\ell \partial_m X^M X^N_{;n} + X^M_{;m}\partial_\ell \partial_n X^N) \; .$$

Permutating the indices ℓ, m, n we may construct the Christoffel symbols of the first kind for the tensor $\underset{\sim}{c}$

$$(4.3) \qquad [\ell m\,,n]_{\underset{\sim}{c}} \equiv \tfrac{1}{2}(\partial_\ell c_{mn} + \partial_m c_{n\ell} - \partial_n c_{\ell m}) =$$

$$= [LM\,,N]_{\underset{\sim}{G}} X^L_{;\ell} X^M_{;m} X^N_{;n} + G_{MN} X^N_{;n}\partial_\ell \partial_m X^M \; ,$$

where $[LM,N]_{\underset{\sim}{G}}$ are the Christoffel symbols for the fundamental tensor $\underset{\sim}{G}$. Since there are 18 equations (4.3); we easily find the derivatives $\partial_\ell \partial_m X^M$:

$$(4.4) \qquad \partial_\ell \partial_m X^N = G^{NK} x^n_{;k}[\ell m\,,n]_{\underset{\sim}{c}} - \begin{Bmatrix} N \\ LM \end{Bmatrix}_{\underset{\sim}{G}} X^L_{;\ell} X^N_{;m} \; .$$

According to (3.18) we have $G^{NK} x^n_{;k} = \overset{-1}{c}{}^{nk} X^N_{;k}$, and since

$$(4.5) \qquad \overset{-1}{c}{}^{nk}[\ell m\,,n]_{\underset{\sim}{c}} = \begin{Bmatrix} k \\ \ell m \end{Bmatrix}_{\underset{\sim}{c}} \; ,$$

(4.4) reduces to

$$(4.6) \qquad \partial_\ell \partial_m X^N = \begin{Bmatrix} n \\ \ell m \end{Bmatrix}_{\underset{\sim}{c}} X^N_{;n} - \begin{Bmatrix} N \\ LM \end{Bmatrix}_{\underset{\sim}{G}} X^L_{;\ell} X^M_{;m} \equiv F^N_{\ell m}.$$

The integrability conditions of (4.6) are $\partial_{[k} F^N_{\ell]m} = 0$.

Differentiation of (4.6) with respect to x^k and the elimination of the second-order derivatives of X's by the aid of (4.6) gives for the integrability conditions the relations

$$R_{k\ell m}^{\cdots n}(\underset{\sim}{c})X^{N}_{;n} - R_{KLM}^{\cdots N}(\underset{\sim}{G})X^{K}_{;k}X^{L}_{;\ell}X^{M}_{;m} = 0$$

where $\underset{\sim}{R}(\underset{\sim}{c})$ and $\underset{\sim}{R}(\underset{\sim}{G})$ are the Riemann-Christoffel tensors (see Appendix, (A4.10)) for the Riemannian connections $\left\{ {k \atop \ell m} \right\}_{\underset{\sim}{c}}$ and $\left\{ {K \atop LM} \right\}_{\underset{\sim}{G}}$. However $\underset{\sim}{G}$ is the metric tensor of Euclidean space and $\underset{\sim}{R}(\underset{\sim}{G})$ vanishes identically. Therefore the integrability conditions reduce to

$$R_{k\ell m}^{\cdots n}(\underset{\sim}{c}) \equiv 2\left(\partial_k \left\{ {n \atop \ell m} \right\}_{\underset{\sim}{c}} + \left\{ {n \atop kt} \right\}_{\underset{\sim}{c}} \left\{ {t \atop \ell m} \right\}_{\underset{\sim}{c}}\right)_{[k\ell]} = 0 . \tag{4.7}$$

Transvecting $R_{k\ell m}^{\cdots t}$ with c_{nt} we obtain the covariant Riemann-Christoffel tensor

$$R_{k\ell mn} = 2\left(\partial_k[\ell m,n]_{\underset{\sim}{c}} - \overset{-1}{c}{}^{st}[\ell m,s]_{\underset{\sim}{c}}[kn,t]_{\underset{\sim}{c}}\right)_{[k\ell]} \tag{4.8}$$

which satisfies the following three identities (cf. Schouten [402]):

$$\begin{aligned} R_{k\ell mn} &= -R_{\ell kmn} , \\ R_{k\ell mn} &= -R_{k\ell nm} , \\ R_{k\ell mn} &= R_{mnk\ell} , \end{aligned} \tag{4.9}$$

and this reduces the number of independent components of the tensor $R_{k\ell mn}$ to six.

The Einstein curvature tensor $\underset{\sim}{A}$ with the components

$$A^{ij} = R^{ij} - \frac{1}{2}R g^{ij} ,$$

where

$$R^{ij} = g^{ki} g^{mj} R_{\ell km}^{\cdot\cdot\cdot\ell} , \qquad R = g_{ij} R^{ij}$$

in three-dimensional spaces may be obtained from (4.8) by

$$A^{ij} = \frac{1}{4}\varepsilon^{ik\ell}\varepsilon^{jmn} R_{k\ell mn} , \tag{4.10}$$

and the compatibility conditions may be expressed in terms of the Einstein tensor, which is symmetric.

The compatibility conditions are usually written in terms of the strain tensor $\underset{\sim}{e}$, and may be derived from (4.8) and (4.10) substituting $\underset{\sim}{c}$ from $(3.20)_2$,

$$\underset{\sim}{c} = \underset{\sim}{g} - 2\underset{\sim}{e}$$

and neglecting the products of the Christoffel symbols in (4.8), as small quantities of the second order. Thus,

$$\varepsilon^{ik\ell}\varepsilon^{jmn} e_{km,\ell n} = 0 \tag{4.11}$$

where "," denotes covariant differentiation with respect to the fundamental tensor $\underset{\sim}{g}$.

If the compatibility conditions (4.8) for a given strain are not satisfied, we may write

$$A^{ij} = \eta^{ij}(\underset{\sim}{e}) \tag{4.12}$$

and $\underset{\sim}{\eta}$ is the incompatibility tensor. In the linearized case we have

$$\eta^{ij} = \varepsilon^{ikl}\varepsilon^{jmn} e_{km,ln} . \qquad (4.13)$$

When $\eta \neq 0$ a deformation of the form (4.1) does not exist and the strain tensor may be interpreted as a tensor which represents a deformation from a non-Euclidean configuration N of the body considered into one of its Euclidean configurations. This interpretation of incompatible strains is applied in the theory of dislocations and in thermoelasticity.

4.1. Incompatible Deformations

When the compatibility conditions are not satisfied but the deformation tensor $\underset{\sim}{C}(\underset{\sim}{X})$, or $\underset{\sim}{c}(\underset{\sim}{x})$ is given, the quantities $X^K_{;k}$ which appear in (3.10),

$$G_{KL} X^K_{;k} X^L_{;l} = c_{kl}(\underset{\sim}{x}) \qquad (4.1.1)$$

are not deformation gradients. In other words, the space with the fundamental tensor $\underset{\sim}{c}$ is not the Euclidean space. To indicate that $X^L_{;l}$ are not deformation gradients (i.e. partial derivatives), we shall introduce the notion of distorsion and denote them by $\theta^L_{(\lambda)}$ and $\theta^{(\lambda)}_L$,

$$\theta^L_{(\lambda)} \theta^{(\lambda)}_M = \delta^L_M , \quad \theta^L_{(\lambda)} \theta^{(\mu)}_L = \delta^\mu_\lambda \qquad (4.1.2)$$

such that

$$\Theta^{(\lambda)}_{L} dX^{L} = du^{\lambda} \tag{4.1.3}$$

and du^{λ} are not coordinates of the Euclidean space. (We may also interprete u^{λ} as non-holonomic coordinates in the Euclidean space). Since $\Theta^{(\lambda)}_{L}$ are not deformation gradients, the Pfaffians (4.1.3) are not integrable and

$$\partial_M \Theta^{(\lambda)}_{L} - \partial_L \Theta^{(\lambda)}_{M} = 2 S_{ML}^{\cdot\cdot K} \Theta^{(\lambda)}_{K} \neq 0 . \tag{4.1.4}$$

To determine the geometry of the non-Euclidean space in which the distorted body is to be now considered, we shall introduce some assumptions: a) The space is a linearly connected space; b) coefficients of linear connection Γ^{K}_{ML} are completely determined by the distorsions; c) the distorsions are smooth and continuously differentiable functions of coordinates X^{K};d) the space admits absolute parallelism and the distorsions represent in it three fields of parallel vectors. From these assumptions we may write

$$\partial_M \Theta^{(\lambda)}_{L} - \Gamma^{N}_{ML} \Theta^{(\lambda)}_{N} = 0 \tag{4.1.5}$$

and the coefficients of connection are determined by the expression

$$\Gamma^{K}_{ML} = \Theta^{K}_{(\lambda)} \partial_M \Theta^{(\lambda)}_{L} . \tag{4.1.6}$$

To bring the body back into the Euclidean space, into its final configuration K_0, we have to subject it to an additional incompatible deformation (distorsion) $\Phi^{(\lambda)}{}_{\ell}$, or $\Phi^{\ell}_{(\lambda)}$, such that

$$\Phi^{\ell}_{(\lambda)} du^{\lambda} = dx^{\ell} . \qquad (4.1.7)$$

Combining the distorsions (4.1.3) and (4.1.7), we obtain

$$dx^{\ell} = \Phi^{\ell}_{(\lambda)} \Theta^{(\lambda)}{}_{L} dX^{L} . \qquad (4.1.8)$$

Since x^m and X^M are coordinates in the Euclidean space, the relation (4.1.8) must be integrable and the products of distorsions $\underset{\sim}{\Phi}_{(\lambda)}$ and $\underset{\sim}{\Theta}^{(\lambda)}$ have to represent deformation gradients,

$$\Phi^{\ell}_{(\lambda)} \Theta^{(\lambda)}{}_{L} = x^{\ell}_{;L} , \qquad (4.1.9)$$

$$\Theta^{L}_{(\lambda)} \Phi^{(\lambda)}{}_{\ell} = X^{L}_{;\ell} . \qquad (4.1.10)$$

It may easily be verified from (4.1.6, 9, 10) that

$$\Gamma^{k}_{m\ell} = \Phi^{k}_{(\lambda)} \partial_m \Phi^{(\lambda)}{}_{\ell} = \Gamma^{K}_{ML} X^{M}_{;m} X^{L}_{;\ell} x^{k}_{;K} + x^{k}_{;K} \frac{\partial^2 X^K}{\partial x^{\ell} \partial x^{m}} . \qquad (4.1.11)$$

If $\overset{*}{g}_{KL}$ and $\overset{*}{g}_{k\ell}$ are the fundamental tensors corresponding to the connection Γ^{K}_{ML} and $\Gamma^{k}_{m\ell}$, respectively, we may define the corresponding strain tensor by

$$(4.1.12) \qquad 2E^*_{KL} = g^*_{KL} - G_{KL} \,, \quad 2e^*_{k\ell} = g_{k\ell} - g^*_{k\ell} \,.$$

It is not possible, however, to determine directly the rate of strain and vorticity tensors. Let x_1^ℓ and x_2^ℓ be two infinitesimally close to one another points in the deformed configuration K,

$$(4.1.13) \qquad x_2^\ell - x_1^\ell = \Delta x^\ell = \Phi^{\ell}_{(\lambda)} du^\lambda \,.$$

Since Δu^λ is determined independently through the difference of coordinates $X_1^K - X_2^K$ in the initial configuration,

$$(4.1.14) \qquad X_2^K - X_1^K = \Delta X^K = \Theta^K_{(\lambda)} du^\lambda ,$$

Δu^λ is independent of time and if the configuration K changes with time, only the distorsion $\underset{\sim}{\Phi}_{(\lambda)}$ may be considered as functions of time. Let the equations of motion of points of the body considered, in its final Euclidean configuration K_0, be $x^k = x^k(t)$. Then

$$(4.1.15) \qquad \dot{x}_2^k - \dot{x}_1^k = \Delta v^k = \dot{\Phi}^k_{(\lambda)} \Delta u^\lambda$$

$$(v^k = \dot{x}^k) \,.$$

But

$$v^k(\underset{\sim}{x}_2) = v^k(\underset{\sim}{x}_1 + \Delta \underset{\sim}{x}) = v^k + v^k_{,\ell} \Delta x^\ell + \ldots ,$$

and we may write

$$\Delta v^k = v^k_{,l}\Delta x^l = \dot{\Phi}^k_{(\lambda)}\Phi^{(\lambda)}_l \Delta x^l . \qquad (4.1.16)$$

Since this relation has to be valid for arbitrary pairs of points $\underset{\sim}{x}_1$ and $\underset{\sim}{x}_2$, the gradients of the velocity vector have to satisfy the relation:

$$v^k_{,l} = \dot{\Phi}^k_{(\lambda)}\Phi^{(\lambda)}_l . \qquad (4.1.17)$$

Using the fundamental tensor g_{mk} of the configuration K, we write

$$v_{m,l} = g_{mk}\dot{\Phi}^k_{(\lambda)}\Phi^{(\lambda)}_l ,$$

and for the rate of strain and for the vorticity tensors we have the expressions

$$d_{ml} = \dot{\Phi}^k_{(\lambda)}g_{k(m}\Phi^{(\lambda)}_{l)} , \quad w_{kl} = \dot{\Phi}^m_{(\lambda)}g_{m[k}\Phi^{(\lambda)}_{l]} . \qquad (4.1.18)$$

From the expression $(4.1.18)_2$ for the vorticity tensor we can calculate its gradients,

$$w_{ml,n} = \left[g_{km}(\dot{\Phi}^k_{(\lambda),n}\Phi^{(\lambda)}_l + \dot{\Phi}^k_{(\lambda)}\Phi^{(\lambda)}_{l,n})\right]_{[ml]} .$$

The distorsions $\underset{\sim}{\Phi}_{(\lambda)}$ are only implicite functions of time, and using (4.1.10) we obtain

$$\dot{\Phi}^k_{(\lambda),n} = (\Phi^k_{(\lambda),j}v^j)_{,n} = \Phi^k_{(\lambda),jn}v^j + \Phi^k_{(\lambda),j}\dot{\Phi}^j_{(\mu)}\Phi^{(\mu)}_n = \qquad (4.1.19)$$

$$= \overline{\Phi^k_{(\lambda),n}}^{\,\cdot} + \Phi^k_{(\lambda),j}\Phi^{(\mu)}_n\dot{\Phi}^j_{(\mu)}$$

and finally,

$$(4.1.20) \quad w_{ml,n} = \Big[g_{km}\big(\Phi^{(\lambda)}_{l}\,\overline{\Phi^{k}_{(\lambda),n}}^{\,\cdot} + \Phi^{(\lambda)}_{l}\,\Phi^{k}_{(\lambda),j}\,\Phi^{(\mu)}_{n}\,\dot{\Phi}^{j}_{(\mu)} + \Phi^{(\lambda)}_{l,n}\,\dot{\Phi}^{k}_{(\lambda)}\big)\Big]_{[ml]} .$$

5. Oriented Bodies

A body to each point of which is assigned a set of vectors $\underset{\sim}{d}_{(\alpha)}, \alpha = 1,2,\ldots,n$, represents an <u>oriented body</u>. The vectors $\underset{\sim}{d}_{(\alpha)}$ are <u>directors</u> of the body. In general, deformations of the directors are independent of the deformations of position.

Let the directors in an undeformed reference configuration K_0 be the vectors

$$(5.1) \qquad \underset{\sim}{D}_{(\alpha)} = \underset{\sim}{D}_{(\alpha)}(\underset{\sim}{X}) ,$$

with the components $\underset{\sim}{D}^{K}_{(\alpha)}$ referred to a material system of reference X^K. A deformation of an oriented body is determined by the equations

$$(5.2) \qquad \underset{\sim}{x} = \underset{\sim}{x}(\underset{\sim}{X})$$

$$\underset{\sim}{d}_{(\alpha)} = \underset{\sim}{d}_{(\alpha)}(\underset{\sim}{D}_{(\alpha)}) = \underset{\sim}{d}_{(\alpha)}(\underset{\sim}{X}) .$$

Directors are not material vectors. For material vectors $\underset{\sim}{D}_{(\alpha)}$ the deformation is determined by the deformation of position,

$$D^k_{(\alpha)} = x^k_{;K} D^K_{(\alpha)} . \qquad (5.3)$$

In an oriented body the vectors

$$\Delta^k_{(\alpha)} = d^k_{(\alpha)} - x^k_{;K} D^K_{(\alpha)} \qquad (5.4)$$

represent the difference between the deformed directors and the vectors obtained from the directors in the reference configuration by the deformation of position.

The Cosserat continuum in the strict sense is a material medium to each point of which there are assigned three directors, which represent rigid triads of unit vectors. The directors in this continuum suffer only rigid rotations, and length and angles between the directors are preserved throughout the motion so that

$$g_{k\ell}\, d^k_{(\alpha)} d^\ell_{(\beta)} = G_{KL} D^K_{(\alpha)} D^L_{(\beta)} = D_{\alpha\beta} = \text{const.} . \qquad (5.5)$$

A medium with deformable directors represents a generalized Cosserat continuum.

5.1 Discrete Systems and Continuum Models

The basic notion in the solid state physics is the crystal lattice. A unit cell of a crystal is composed of four lattice points M_0, M_1, M_2, M_3. Let M_0 be a lattice point. Any three vectors $\underset{\sim}{a}_1, \underset{\sim}{a}_2, \underset{\sim}{a}_3$ are lattice vectors if they are position vectors of the lattice points M_1, M_2, M_3 with respect to M_0 of the unit cell. The vectors

$$(5.1.1) \qquad \underset{\sim}{r} = \ell \underset{\sim}{a}_1 + m \underset{\sim}{a}_2 + n \underset{\sim}{a}_3 \quad (\ell, m, n \text{ – integral numbers})$$

determine the lattice points of a perfect crystal.

Motions of a crystal are determined if determined are the motions of its lattice points. However, instead of the motions of the lattice points it is possible to regard the motions of one lattice point for each cell, and the motions of the lattice vectors $\underset{\sim}{a}_\lambda$ for each individual cell. This may be considered as a four-point model which under suitable assumptions may be used for a continuum approximation of an oriented body, as was done by Stojanović, Djurić and Vujosević [428] . A more general approach to the generalized Cosserat continuum with an arbitrary number of directors is proposed by Rivlin [377, 378] and in the following we shall consider Rivlin's n-point model.

We assume that a body consists of <u>particles</u> $P_1, \ldots, P_N$ and that each particle consists of n <u>material points</u> $M_1, \ldots, M_n$ with masses $m_1, \ldots m_n$ and with position vectors $\underset{\sim}{r}_1, \ldots, \underset{\sim}{r}_n$ with respect to a fixed origin $\underset{\sim}{O}$ in the space.

If C_P is the centre of masses of the particle P, and $\underset{\sim}{\varrho}_\nu$, $\nu = 1, \ldots, n$ position vectors of the points M_ν, from particle dynamics we obtain for the momentum, moment of momentum and kinetic energy of a particle P the following expressions:*

*Rivlin [377, 378] investigated the transition from a discrete ./.

$$\underset{\sim}{K} = \sum_{\nu=1}^{n} m_\nu \dot{\underset{\sim}{r}}_\nu = \sum_{\nu=1}^{n} m_\nu \underset{\sim}{v}_\nu \ , \tag{5.1.2}$$

$$\underset{\sim}{\ell}^0 = \sum_{\nu=1}^{n} m_\nu \underset{\sim}{r}_\nu \times \underset{\sim}{v}_\nu = m \underset{\sim}{r}_c \times \underset{\sim}{v}_c + \sum_{\nu=1}^{n} m_\nu \underset{\sim}{\varrho}_\nu \times \dot{\underset{\sim}{\varrho}}_\nu \ , \tag{5.1.3}$$

$$T = \frac{1}{2}\left(m v_c^2 + \sum_{\nu=1}^{n} m_\nu \dot{\underset{\sim}{\varrho}}_\nu \cdot \dot{\underset{\sim}{\varrho}}_\nu \right) . \tag{5.1.4}$$

Here we have

$$\underset{\sim}{v}_\nu = \dot{\underset{\sim}{r}}_\nu = \frac{\partial \underset{\sim}{r}_\nu}{\partial t} \ , \tag{5.1.5}$$

$$\underset{\sim}{\varrho}_\nu = \underset{\sim}{r}_\nu - \underset{\sim}{r}_c \ , \tag{5.1.6}$$

$$\sum_{\nu=1}^{n} m_\nu \underset{\sim}{\varrho}_\nu = 0 \ , \tag{5.1.7}$$

$$m = \sum_{\nu=1}^{n} m_\nu \ , \tag{5.1.8}$$

$$m \underset{\sim}{r}_c = \sum_{\nu=1}^{n} m_\nu \underset{\sim}{r}_\nu \ . \tag{5.1.9}$$

Introducing the coefficients (which are not tensors)

$$i^{\lambda\mu} = \frac{1}{m} \sum_{\nu=1}^{n} m_\nu \delta_\nu^\lambda \delta_\nu^\mu \ , \tag{5.1.10}$$

---

system to continuum, including some implications of the first and second laws of thermodynamics, without writing the expressions for momentum and moment of momentum.

the relations (5.1.3, 4) may be rewritten in the form

$$\underset{\sim}{l}^0 = m(\underset{\sim}{r}_c \times \underset{\sim}{v}_c + \underset{\sim}{i}^{\lambda\mu} \underset{\sim}{\rho}_\lambda \times \dot{\underset{\sim}{\rho}}_\mu), \tag{5.1.11}$$

$$T = \frac{m}{2}(v_c^2 + i^{\lambda\mu} \dot{\underset{\sim}{\rho}}_\lambda \cdot \dot{\underset{\sim}{\rho}}_\mu). \tag{5.1.12}$$

From the last two expressions we see that for the dynamical specification of the particle P we need to know the quantities: m – the mass of the particle, $i^{\lambda\mu}$ – the dimensionless coefficients which characterize the distribution of masses inside the particle, and the vectors $\underset{\sim}{\rho}_\lambda$ which determine the configuration of the particle.

To denote that all the quantities which appear in (5.1.2 - 12) correspond to the particle P we shall label them with the index P so that we write

$$\underset{\sim}{K}_P, \quad \underset{\sim}{l}_P^0, \quad m_P, \quad m_\nu^P, \quad \underset{\sim}{r}_\nu^P, \quad \underset{\sim}{r}_c^P, \quad \underset{\sim}{\rho}_\nu^P, \quad \dot{\underset{\sim}{\rho}}_\nu^P, \quad T_P,$$

and

$$m_P = \sum_{\nu=1}^{n} m_\nu^P, \quad m_P \underset{\sim}{r}_c^P = \sum_{\nu=1}^{n} m_\nu^P \underset{\sim}{r}_\nu^P, \tag{5.1.13}$$
$$\underset{\sim}{K}_P = \sum_{\nu=1}^{n} m_\nu^P \dot{\underset{\sim}{r}}_\nu^P, \quad \underset{\sim}{l}_P^0 = m_P(\underset{\sim}{r}_c^P \times \dot{\underset{\sim}{r}}_c^P + i_P^{\lambda\mu} \underset{\sim}{\rho}_\lambda^P \times \dot{\underset{\sim}{\rho}}_\mu^P).$$

For a body consisting of N particles we have now for the momentum

$$K = \sum_{P=1}^{N} m_P \dot{\underset{\sim}{r}}_c^P, \tag{5.1.14}$$

for the moment of momentum

$$\underset{\sim}{l}^{0} = \sum_{P=1}^{N} \underset{\sim}{l}_{P}^{0} = \sum_{P=1}^{N} m_P \underset{\sim}{r}_c^P \times \dot{\underset{\sim}{r}}_c^P + \sum_{P=1}^{N} m_P i_P^{\lambda\mu} \underset{\sim}{\varrho}_\lambda^P \times \dot{\underset{\sim}{\varrho}}_\mu^P , \quad (5.1.15)$$

and for the kinetic energy

$$T = \sum_{P=1}^{N} T_P = \frac{1}{2}\sum_{P=1}^{N} m_P (\dot{\underset{\sim}{r}}_c^P \cdot \dot{\underset{\sim}{r}}_c^P + i_P^{\lambda\mu} \dot{\underset{\sim}{\varrho}}_\lambda^P \cdot \dot{\underset{\sim}{\varrho}}_\mu^P) . \quad (5.1.16)$$

To pass from this discrete system of particles to a continuum we have to replace the sums by integrals. In order to do so we assume that our system of particles occupies a domain $B + \partial B$ where ∂B is the boundary of the body B . We assume further that the discrete vectors $\underset{\sim}{r}_c^P$, $\dot{\underset{\sim}{r}}_c^P$, $\underset{\sim}{\varrho}_\nu^P$ and $\dot{\underset{\sim}{\varrho}}_\nu^P$ may be replaced by continuous vector fields $\underset{\sim}{r}$, $\dot{\underset{\sim}{r}}$ and $\underset{\sim}{d}_{(\nu)}$ and $\dot{\underset{\sim}{d}}_{(\nu)}$ and the discrete scalars m_P and $i_P^{\lambda\mu}$ by continuous scalar fields ϱ and $i^{\lambda\mu}$. It must be noted that the passage from a system of particles to a continuous model can be effected only if all the quantities involved, which are connected with the particles, vary but little as we pass from one particle to its neighbours.

We assume that a region V of B with a boundary S is sufficiently large to contain many particles. Hence we may write

$$\sum_V m_P = \int_V \varrho \, dV , \quad (5.1.17)$$

$$\sum_V m_\nu^P = \int_V \varrho_\nu \, dV , \quad (5.1.18)$$

$$\sum_V m_P \underset{\sim}{r}{}^P_c = \int_V \varrho \underset{\sim}{r}\, dV , \tag{5.1.19}$$

$$\sum_V m_P \underset{\sim}{r}{}^P_c \times \dot{\underset{\sim}{r}}{}^P_c = \int_V \varrho \underset{\sim}{r} \times \dot{\underset{\sim}{r}}\, dV , \tag{5.1.20}$$

$$\sum_V m_P \dot{\underset{\sim}{r}}{}^P_c \cdot \dot{\underset{\sim}{r}}{}^P_c = \int_V \varrho \dot{\underset{\sim}{r}} \cdot \dot{\underset{\sim}{r}}\, dV , \tag{5.1.21}$$

$$\sum_V m_P i^{\lambda\mu}_P \underset{\sim}{\varrho}{}^P_\lambda \times \dot{\underset{\sim}{\varrho}}{}^P_\mu = \int_V \varrho i^{\lambda\mu} \underset{\sim}{d}_{(\lambda)} \times \dot{\underset{\sim}{d}}_{(\mu)} dV , \tag{5.1.22}$$

$$\sum_V m_P i^{\lambda\mu}_P \dot{\underset{\sim}{\varrho}}{}^P_\lambda \cdot \dot{\underset{\sim}{\varrho}}{}^P_\mu = \int_V \varrho i^{\lambda\mu} \dot{\underset{\sim}{d}}_{(\lambda)} \cdot \dot{\underset{\sim}{d}}_{(\mu)} dV . \tag{5.1.23}$$

Thus the expressions for momentum, moment of momentum and for the kinetic energy for a part V of the body B obtain the form

$$\underset{\sim}{K} = \int_V \varrho \dot{\underset{\sim}{r}}\, dV , \tag{5.1.24}$$

$$\underset{\sim}{\ell}{}^0 = \int_V \varrho (\underset{\sim}{r} \times \dot{\underset{\sim}{r}} + i^{\lambda\mu} \underset{\sim}{d}_{(\lambda)} \times \dot{\underset{\sim}{d}}_{(\mu)}) dV , \tag{5.1.25}$$

$$T = \frac{1}{2} \int_V \varrho (\dot{\underset{\sim}{r}} \cdot \dot{\underset{\sim}{r}} + i^{\lambda\mu} \dot{\underset{\sim}{d}}_{(\lambda)} \cdot \dot{\underset{\sim}{d}}_{(\mu)}) dV . \tag{5.1.26}$$

The continuum representation of the originally discrete system has all the properties of a generalized Cosserat medium: to its points r attached are the directors $\underset{\sim}{d}_{(\lambda)}$, the motions of which are independent of the motions of the points.

5.2 Materials with Microstructure

Let a body be composed of microelements $\Delta V'$ in which a continuous mass density ρ' exists, such that the microelements $\Delta V'$ represent material continua. A macro-volume element dV is composed of the micro-volume elements dV',

$$dV = \int_{dV} dV' , \tag{5.2.1}$$

and we assume that the macro-mass dM in dV is the average of all masses in dV. Denoting by $\rho' dV' = dM'$ the micro-mass of the micro-volume element dV', we may write

$$\int_{dV} \rho' dV' = dM = \rho dV . \tag{5.2.2}$$

With respect to a fixed Cartesian coordinate system Z^{α} let Z'^{α} be coordinates of points $\underset{\sim}{Z}'$ in a micro-volume element dV' in a reference configuration K_0. The integral over the macro-volume element

$$\int_{dV} \rho' Z'^{\alpha} dV' = \rho Z^{\alpha} dV \tag{5.2.3}$$

determines the centre of mass $\underset{\sim}{Z}$ of the macro-volume element dV. Denoting by $\underset{\sim}{R}' = Z'^{\alpha} \underset{\sim}{e}_{\alpha}$ the position vectors of the points $\underset{\sim}{Z}'$ of microelements, by $\underset{\sim}{R} = Z^{\alpha} \underset{\sim}{e}_{\alpha}$ the position vectors of the centres of mass of macro-volume elements dV and by $\underset{\sim}{P}' = \Xi'^{\alpha} \underset{\sim}{e}_{\alpha}$ the position vectors of the points $\underset{\sim}{R}'$ relative to the centre of gravity $\underset{\sim}{R}$,

$$\underset{\sim}{R}' = \underset{\sim}{R} + \underset{\sim}{P}' , \tag{5.2.4}$$

all with respect to a fixed Cartesian system of reference, we have in the coordinate notation

$$Z'^{\alpha} = Z^{\alpha} + \Xi'^{\alpha} . \tag{5.2.5}$$

In a deformed configuration $K(t)$ let the positions of points $\underset{\sim}{R}'$ be $\underset{\sim}{r}'$ and of the points $\underset{\sim}{R}$ be $\underset{\sim}{r}$. The relative position vectors of $\underset{\sim}{r}'$ with respect to the new positions of the centres of mass let be $\underset{\sim}{\rho}'$. The equations of motion of the centres of mass of the macro-elements dV, which become dv and of the points $\underset{\sim}{R}'$ are

$$\underset{\sim}{r} = \underset{\sim}{r}(\underset{\sim}{R},t) , \quad \underset{\sim}{R} = \underset{\sim}{R}(\underset{\sim}{r},t) , \tag{5.2.6}$$

$$\underset{\sim}{r}' = \underset{\sim}{r}(\underset{\sim}{R}',t) , \quad \underset{\sim}{R}' = \underset{\sim}{R}(\underset{\sim}{r}',t) ,$$

and we assume that in the deformed configuration the positions of the points $\underset{\sim}{Z}'$ are defined by the relations

$$\underset{\sim}{r}' = \underset{\sim}{r} + \underset{\sim}{\rho}' , \quad \text{or} \quad z'^{\alpha} = z^{\alpha} + \xi'^{\alpha} . \tag{5.2.7}$$

The further assumption we make is that the motion (5.2.6) carries the centres of mass of dV into the centres of mass of the deformed macro-volume elements dv,

$$\int_{dv} \rho' \underset{\sim}{r}' dv' = \rho \underset{\sim}{r} dv . \tag{5.2.8}$$

From (5.2.6) we have

$$\underset{\sim}{r}' = \underset{\sim}{r}(\underset{\sim}{R} + \underset{\sim}{P}',t) = \underset{\sim}{r}(\underset{\sim}{R},t) + \underset{\sim}{\varrho}' , \tag{5.2.9}$$

where

$$\underset{\sim}{\varrho}' = \underset{\sim}{\varrho}(\underset{\sim}{R},\underset{\sim}{P}',t) . \tag{5.2.10}$$

Expanding $(5.2.9)_1$, under the assumption that $\underset{\sim}{\varrho}'$ is an analytic function of Ξ'^{α}, we obtain

$$\underset{\sim}{\varrho}' = \underset{\sim}{\varrho}(\underset{\sim}{R},0,t) + \frac{\partial \underset{\sim}{\varrho}^{\gamma}}{\partial \Xi'^{\alpha}} \Xi'^{\alpha} + \ldots \tag{5.2.11}$$

Through $(5.2.9)_2$ we see that for $\underset{\sim}{P}' = 0$

$$\underset{\sim}{\varrho}(\underset{\sim}{R},0,t) = \underset{\sim}{0} , \tag{5.2.12}$$

and if we write

$$\frac{\partial \underset{\sim}{\varrho}}{\partial \Xi'^{\alpha}} \equiv \chi_{\alpha}(\underset{\sim}{R},t) , \quad \chi^{\beta}_{\cdot\alpha} = \frac{\partial \xi^{\beta}}{\partial \Xi'^{\alpha}} , \tag{5.2.13}$$

in the linear approximation we obtain the equations of motion of points $\underset{\sim}{R}'$ in the form

$$\underset{\sim}{\varrho}' = \underset{\sim}{\chi}_{\alpha} \Xi'^{\alpha} , \tag{5.2.14}$$

or

$$\xi'^{\lambda} = \chi^{\lambda}_{\cdot\alpha} \Xi'^{\alpha} . \tag{5.2.15}$$

The coefficients $\underset{\sim}{\chi}^{\alpha}$ reciprocal to $\underset{\sim}{\chi}_{\alpha}$ are defined by the relations

$$(5.2.16) \qquad \chi_{\beta}^{\cdot\alpha} = \frac{\partial \Xi'^{\alpha}}{\partial \xi'^{\beta}} ,$$

and

$$(5.2.17) \qquad \chi^{\alpha}_{\cdot\beta}\chi^{\cdot\gamma}_{\alpha} = \delta^{\gamma}_{\beta} , \quad \chi^{\alpha}_{\cdot\beta}\chi^{\cdot\beta}_{\gamma} = \delta^{\alpha}_{\gamma} .$$

The velocity $\underset{\sim}{v}'$ of a point $\underset{\sim}{R}'$ is defined by

$$(5.2.18) \qquad \underset{\sim}{v}' = \dot{\underset{\sim}{r}} = \dot{\underset{\sim}{r}} + \dot{\underset{\sim}{\varrho}}' = \underset{\sim}{v} + \dot{\underset{\sim}{\chi}}_{\alpha}\Xi'^{\alpha}$$

or, in the componental form

$$(5.2.19) \qquad \dot{z}'^{\alpha} = \dot{z}^{\alpha} + \dot{\chi}^{\alpha}_{\cdot\beta}\Xi'^{\alpha} .$$

Eliminating Ξ'^{β} from (5.2.19) we obtain

$$(5.2.20) \qquad v'^{\alpha} = v^{\alpha} + \dot{\chi}^{\alpha}_{\cdot\beta}\chi^{\cdot\beta}_{\gamma}\xi'^{\gamma} = v^{\alpha} + \nu^{\alpha}_{\cdot\gamma}\xi'^{\gamma} ,$$

where

$$(5.2.21) \qquad \nu^{\alpha}_{\cdot\gamma} = \dot{\chi}^{\alpha}_{\cdot\beta}\chi^{\cdot\beta}_{\gamma} = \nu^{\alpha}_{\cdot\gamma}[\underset{\sim}{z}(\underset{\sim}{z},t),t] = \nu^{\alpha}_{\cdot\gamma}(z,t) .$$

For a macro-volume element dv the momentum is giv-en by the relation

$$(5.1.22) \qquad d\underset{\sim}{K} = \int_{dv} d\underset{\sim}{K}' = \int_{dv} \varrho'\underset{\sim}{v}'dv' = \int_{dv} \varrho'(\underset{\sim}{v} + \nu^{\alpha}_{\cdot\beta}\xi'^{\beta})dv' = \varrho\underset{\sim}{v}\,dv ,$$

and for a portion v of a body we have

$$(5.2.23) \qquad \underset{\sim}{K} = \int_{v} \varrho\underset{\sim}{v}\,dv .$$

The moment of momentum $d\underset{\sim}{\ell}^0$ for the macro-volume element dv will be

$$d\underset{\sim}{\ell}^0 = \int_{dv} \varrho' \underset{\sim}{r}' \times \underset{\sim}{v}' dv' = \int_{dv} \varrho'(\underset{\sim}{r} + \underset{\sim}{\varrho}') \times (\underset{\sim}{v} + \dot{\underset{\sim}{x}}_\alpha \Xi'^\alpha) dv' . \quad (5.2.24)$$

Since $\underset{\sim}{\varrho}'$ are the position vectors of the points $\underset{\sim}{r}'$ relative to the centre of mass $\underset{\sim}{r}$ of the macro-volume element, we have

$$\int_{dv} \varrho' \underset{\sim}{\varrho}' dv' = 0 ,$$

and

$$d\underset{\sim}{\ell}^0 = \varrho \underset{\sim}{r} \times \underset{\sim}{v} dv + \int_{dv} \varrho' \underset{\sim}{\varrho}' \times \dot{\underset{\sim}{x}}_\alpha \Xi'^\alpha dv' . \quad (5.2.25)$$

In the componental form we have

$$\underset{\sim}{\varrho}' \times \dot{\underset{\sim}{x}}_\alpha \Xi'^\alpha = \varepsilon_{\lambda\mu\nu} \xi'^\lambda \dot{x}^\mu_{.\alpha} \Xi'^\alpha \underset{\sim}{e}^\nu , \quad (5.2.26)$$

and using (5.2.14) this becomes

$$\underset{\sim}{\varrho}' \times \dot{\underset{\sim}{x}}_\alpha \Xi'^\alpha = \varepsilon_{\lambda\mu\nu} x^\lambda_{.\beta} \dot{x}^\mu_{.\alpha} \Xi'^\alpha \Xi'^\beta \underset{\sim}{e}^\nu = \quad (5.2.27)$$

$$= \underset{\sim}{x}_\beta \times \dot{\underset{\sim}{x}}_\alpha \Xi'^\alpha \Xi'^\beta .$$

Hence, for the moment of momentum $d\underset{\sim}{\ell}^0$ we may write

$$d\underset{\sim}{\ell}^0 = \varrho \underset{\sim}{r} \times \underset{\sim}{v} dv + \underset{\sim}{x}_\alpha \times \dot{\underset{\sim}{x}}_\beta \int_{dv} \varrho' \Xi'^\alpha \Xi'^\beta dv' . \quad (5.2.28)$$

Using the inverse of (5.2.15),

$$\Xi'^\alpha = x_\lambda^{.\alpha} \xi'^\lambda , \quad (5.2.29)$$

by (5.2.13) we see that

$$(5.2.30) \qquad \int_{dv} \varrho' \Xi'^{\alpha} \Xi'^{\beta} dv' = x_{;\lambda}^{\cdot\alpha} x_{;\mu}^{\cdot\beta} \int_{dv} \varrho' \xi'^{\lambda} \xi'^{\mu} dv' ,$$

and if we introduce the "micro-inertia density", $i^{\lambda\mu}$ by the expression

$$(5.2.31) \qquad \varrho i^{\lambda\mu} dv = \int_{dv} \varrho' \xi'^{\lambda} \xi'^{\mu} dv' ,$$

and the "macro-inertia density moments" $I^{\alpha\beta}$ by

$$(5.2.32) \qquad I^{\alpha\beta} = x_{;\lambda}^{\cdot\alpha} x_{;\mu}^{\cdot\beta} i^{\lambda\mu} ,$$

the expression (5.2.28) for the moment of momentum becomes

$$(5.2.33) \qquad d\underset{\sim}{\ell}^{0} = \varrho \underset{\sim}{r} \times \underset{\sim}{v} + \varrho i^{\lambda\mu} \underset{\sim}{x}_{\alpha} \times \dot{\underset{\sim}{x}}_{\mu} dv .$$

For a portion v of the body we have now

$$(5.2.34) \qquad \underset{\sim}{\ell}^{0} = \int_{v} d\underset{\sim}{\ell}^{0} = \int_{v} \varrho(\underset{\sim}{r} \times \underset{\sim}{v} + i^{\lambda\mu} \underset{\sim}{x}_{\lambda} \times \dot{\underset{\sim}{x}}_{\mu}) dv .$$

Analogously, we find for the kinetic energy the expression

$$(5.2.35) \qquad T = \frac{1}{2} \int_{v} \varrho(\underset{\sim}{v} \cdot \underset{\sim}{v} + i^{\lambda\mu} \dot{\underset{\sim}{x}}_{\lambda} \cdot \dot{\underset{\sim}{x}}_{\mu}) dv .$$

Materials with micro-structure were first considered by Eringen and Suhubi in elasticity [138, 442] and in the fluid mechanics [124] . Here we diverged slightly from the original exposition of Eringen and Suhubi since we wanted to write the expressions for $\underset{\sim}{\ell}^{0}$ and T in a form similar to the corresponding formula in the section 5.1, obtained from the consideration of a discrete system.

In the original papers (cf. [124]) the coefficients $I^{\alpha\beta}$ stay instead of $i^{\alpha\beta}$, and $i^{\alpha\beta}$ instead of $I^{\alpha\beta}$, and, following our notation, the coefficients

$$I^{\lambda\mu} = x_{;\alpha}^{\cdot\lambda} x_{;\beta}^{\cdot\mu} \int_{dV} P' \Xi'^{\alpha} \Xi'^{\beta} dV' \tag{5.2.36}$$

are named "micro-inertia moments", and the coefficients

$$i^{\alpha\beta} = \int_{dV} P' \Xi'^{\alpha} \Xi'^{\beta} dV' \tag{5.2.37}$$

are constant material coefficients. We prefer to use here the densities defined by (5.2.31, 32)

According to Eringen [123], materials affected by micro-motion and micro-deformation are micromorphic materials.

Micropolar media are a subclass of micromorphic materials, and they exhibit microrotational effects, i.e. the material points in a volume element can undergo only the rotational motions about the centres of mass.

The materials with microstructure of Mindlin [285, 289] coincide with the model given above. Mindlin considered the infinitesimal deformations only, and his theory is restricted to the linear case. If we assume that the deformations are infinitesimal and if we make no distinction between the material and spatial coordinates Z^{α} and z^{α}, for the micro- deformation we may write

(5.2.38) $$\xi'^{\beta} = \Xi'^{\beta} + u'^{\beta} ,$$

where u'^{α} are components of the micro-displacements. From (5.2.15) it follows then

(5.2.39) $$u'^{\beta} = (x^{;\beta}_{\lambda} - \delta^{\beta}_{\lambda})\Xi'^{\lambda} \approx (x^{;\beta}_{\lambda} - \delta^{\beta}_{\lambda})\xi'^{\lambda} ,$$

where the quantities $\psi^{\cdot\beta}_{\lambda}$ defined by the expression

(5.2.40) $$\psi^{\cdot\beta}_{\lambda} = \frac{\partial u'^{\beta}}{\partial \xi'^{\lambda}} = x^{;\beta}_{\lambda} - \delta^{\beta}_{\lambda}$$

are called by Mindlin the <u>micro-deformations</u>. Denoting by u^{α} the displacements of particles (which are not necessarily represented by their centres of mass),

(5.2.41) $$u^{\alpha} = z^{\alpha} - Z^{\alpha}$$

the <u>macro-strain</u> is given by

(5.2.42) $$\varepsilon_{\alpha\beta} = \frac{1}{2}\left(\frac{\partial u_{\beta}}{\partial z^{\alpha}} + \frac{\partial u_{\alpha}}{\partial z^{\beta}}\right) ,$$

and the <u>relative deformation</u> by

(5.2.43) $$\gamma_{\alpha\beta} = \frac{\partial u_{\beta}}{\partial z^{\alpha}} - \psi_{\alpha\beta} .$$

In this theory the quantities $\psi_{\alpha\beta}$ play the role of directors, and the medium with micro-structure is a generalized Cosserat medium.

5.3. Multipolar Theories

In a series of papers Green and Rivlin [172, 173, 175] , Green [151] , and Green Naghdi and Rivlin [170] developed the theory of multipolar continua which represents a very general, but a very formal approach. Let Z^{α} be coordinates of a particle in reference position and z^{α} its position at time

$$z^{\alpha}(\tau) = z_{\alpha}(\underset{\sim}{Z},\tau) , \quad -\infty < \tau \leq t . \tag{5.3.1}$$

It is possible to consider the position of the particle $\underset{\sim}{Z}$ at time τ also in terms of the current position at time t, so that

$$z^{\alpha}(\tau) = z^{\alpha}(z^1, z^2, z^3, \tau, t) . \tag{5.3.2}$$

A simple 2^{γ}-pole displacement field is defined in two forms,

$$z_{\alpha B_1 \ldots B_{\gamma}}(\tau) = z_{\alpha B_1 \ldots B_{\gamma}}(Z , \tau) , \tag{5.3.3}$$

and

$$z_{\alpha\beta_1 \ldots \beta_{\gamma}}(\tau) = z_{\alpha\beta_1 \ldots \beta_{\gamma}}(\underset{\sim}{z} , t , \tau) . \tag{5.3.4}$$

The examples of such multipolar displacement fields are the gradients

$$z^{\alpha}_{.B_1 \ldots B_{\gamma}}(\tau) = \frac{\partial^{\gamma} z^{\alpha}(\tau)}{\partial Z^{B_1} \ldots \partial Z^{B_{\gamma}}} \tag{5.3.5}$$

and

$$z^{\alpha}_{\cdot\beta_1\ldots\beta_\gamma}(\tau) = \frac{\partial^\gamma z^\alpha(\tau)}{\partial z^{\beta_1}\ldots\partial z^{\beta_\gamma}} . \tag{5.3.6}$$

The time derivatives of the multipolar displacements represent the multipolar (2^γ-pole) velocity fields.

In multipolar theories the deformation is describ_ ed by the simple deformation field $\underset{\sim}{z}(\tau)$ and by ν tensor fields, say $u_{\alpha\Lambda_1\ldots\Lambda_\gamma}(\tau)$, $\gamma = 1,2,\ldots,\nu$. The tensor fields $u_{\alpha\Lambda_1\ldots\Lambda_\gamma}(\tau)$ are called multipolar deformation fields. In 1967 Green and Rivlin [175] showed that the multipolar theory can be considered as a special case of the director theory, with the multipolar deformation fields $u_{\alpha\Lambda_1\ldots\Lambda_\gamma}$ corresponding to 3^γ directors.

The theory of multipolar media was applied by Bleustein and Green to fluids [38].

5.4 Strain-Gradient Theories

The state of strain of a body at a point $\underset{\sim}{X}$ depends on the relative displacements of points in a neighbourhood $N(\underset{\sim}{X})$. If $\underset{\sim}{X} + \Delta\underset{\sim}{X}$ is a point in $N(\underset{\sim}{X})$, and the equations of motion are

$$x^i = x^i(\underset{\sim}{X},t) \tag{5.4.1}$$

the relative displacements of all points $\underset{\sim}{X}+\Delta\underset{\sim}{X}$ for arbitrary $\Delta\underset{\sim}{X}$ are determined by the deformation gradients

$$x^i_{;K},\ x^i_{;K_1K_2},\ldots,\ x^i_{;K_1\ldots K_N},\ldots\ . \tag{5.4.2}$$

Material derivatives of these deformation gradients are the velocity gradients,

$$v^i_{;K},\ v^i_{;K_1K_2},\ldots,\ v^i_{;K_1\ldots K_N},\ldots\ . \tag{5.4.3}$$

The theories which consider the influence of the higher-order deformation and velocity gradients are known as the strain-gradient theories.

According to (3.20) and (3.11), by differentiation we obtain

$$E_{KL,M} = g_{k\ell}x^k_{;M(L}x^\ell_{;K)}\ , \tag{5.4.4}$$

and we see that the first gradient of strain involves the second gradient of deformation.

The deformed directors at two points, say $\underset{\sim}{X}$ and $\underset{\sim}{X} + \Delta\underset{\sim}{X}$ in a neighbourhood $N(\underset{\sim}{X})$ will be according to (5.2)

$$d^k_{(\alpha)} = d^k_{(\alpha)}(\underset{\sim}{X})\ , \tag{5.4.5}$$

$$d^k_{(\alpha)}(\underset{\sim}{X} + \Delta\underset{\sim}{X}) = d^k_{(\alpha)}(\underset{\sim}{X}) + d^k_{(\alpha);L}\Delta X^L + \ldots\ .$$

Hence, the director deformation at $\underset{\sim}{X}$ is characterized by the director gradients $d^k_{(\alpha);L}, d^k_{(\alpha);L_1L_2},\ldots$. From (5.4) it follows then that

$$d^k_{(\alpha);L} = \Delta^k_{(\alpha);L} + x^k_{;KL}D^K_{(\alpha)} + x^k_{;K}D^K_{(\alpha);L}\ . \tag{5.4.6}$$

If an oriented body degenerates into an ordinary body the directors will become material vectors and $\Delta^k_{(\alpha)}$ vanishes. In this case we may choose the directors $\underset{\sim}{D}_{(\alpha)}$ in the reference configuration to be parallel vector fields so that $D^k_{(\alpha);L} = 0$. Consequently, the director gradients will be proportional to the second gradients of deformation,

$$d^k_{(\alpha);L} = x^k_{;KL} D^K_{(\alpha)} , \tag{5.4.7}$$

and the theory of an oriented body will degenerate into a strain--gradient theory.

In Cosserat bodies the directors $\underset{\sim}{d}_{(\alpha)}$ form rigid triads, such that

$$\underset{\sim}{d}_{(\alpha)} \cdot \underset{\sim}{d}_{(\beta)} = \underset{\sim}{D}_{(\alpha)} \cdot \underset{\sim}{D}_{(\beta)} = \text{const.}. \tag{5.4.8}$$

In this case the rates of the directors will be

$$\dot{\underset{\sim}{d}}_{(\alpha)} = \underset{\sim}{\omega} \times \underset{\sim}{d}_{(\alpha)} , \tag{5.4.9}$$

where $\underset{\sim}{\omega}$ is the rate of rotation of the triads of directors. In the componental form we may write

$$\dot{\underset{\sim}{d}}_{(\alpha)m} = \varepsilon_{mij} \omega^i d^j_{(\alpha)} = \omega_{jm} d^j_{(\alpha)} . \tag{5.4.10}$$

If there are only three directors, $\alpha = 1,2,3$ and, in the Cosserat continuum in the strict sense there are only three directors, the reciprocal triads $\underset{\sim}{d}^{(\alpha)}$ exist, and for the tensor $\underset{\sim}{\omega}$ we have

$$\omega_{nm} = d^{(\alpha)}_n \dot{d}_{(\alpha)m} . \tag{5.4.11}$$

From (5.4.8) it follows that the left-hand side of (5.4.11) is an antisymmetric tensor. If the rotations of the director triads are constrained to follow the rotations of the medium, which are given by

$$\omega_{nm} = v_{[n,m]} , \qquad (5.4.12)$$

where $v^i = \dot{x}^i$ is the velocity vector, for the corresponding medium it is said that it is a Cosserat continuum with constrained rotations (Toupin [463]).

5.5 Shells and Rods as Oriented Bodies

One of the essential problems in the theory of structures is the simplification of the general three-dimensional theories of materials. All structures are three-dimensional bodies, but certain geometric properties justify the introduction of approximations which give sufficiently good results, at least for engineering purposes. In the Introduction to these lecture notes we mentioned St. Venant's remark that for the description of thin bodies an analysis of deformation of a straight line, or of a surface, is insufficient, but an extensible line may serve as the first approximation for a rod. Deformable planes and surfaces play the same role in the theory of plates and shells. The main question is what is happening with the points which in an initial configuration were situated outside the middle surface of the shell considered, or which were not on the middle line of a rod. There is a number of different hypothesis (Kirchoff, Love, Vlasov etc.), and all these hypothesis have a definite value, under corresponding assumptions.

In this section we shall give a brief account of the approximations of a three-dimensional medium for shells and rods, according to the theory recently developed by Green, and Naghdi [157] and by Green and Naghdi [169].

5.5.1 Shells

Let $X^3=0$ define a surface $\underline{S}$ in the initial configuration of a body, and let the position vector of any point on $\underline{S}$ be

$$\underset{\sim}{R} = \underset{\sim}{R}(X^1, X^2) . \tag{5.5.1.1}$$

For the surface $\underline{S}$ we assume that it is smooth and non-intersecting. At time t the surface $\underline{S}$ will he $\underline{s}$ and the points on $\underline{s}$ are determined by the position vector

$$\underset{\sim}{r} = \underset{\sim}{r}(X^1, X^2, t) . \tag{5.5.1.2}$$

We further assume that a three-dimensional body is bounded by the surfaces

$$X^3 = A , \; X^3 = B \qquad (A < 0 < B) \tag{5.5.1.3}$$

and by a surface

$$F(X^1,X^2) = 0 . \qquad (5.5.1.4)$$

The relations (5.5.1.3, 4) fix a shell in the initial (reference) configuration.

For the simplicity in writing we shall put $X^3 = X$, and we shall let the Greek indices take the values 1, 2.

At time t we may introduce spatial coordinates x^k such that

$$x^\alpha = x^\alpha(X^1,X^2;t)$$

$$x^3 \equiv x = x(X^1,X^2,X;t) ,$$

and we assume that the shell in this instant of time is fixed by the bounding surfaces

$$x = \alpha , \quad x = \beta , \quad (\alpha < 0 < \beta)$$

$$f(x^1,x^2) = 0 .$$

The coordinates x^k may be selected to be convected coordinates and then we have

$$x^k = \delta^k_K X^K .$$

The position vector of any point of the shell is a function of coordinates and time,

(5.5.1.5) $$\underset{\sim}{r}^* = \underset{\sim}{r}^*(X^1, X^2, X; t) ,$$

and for sufficiently small α and β we may represent $\underset{\sim}{r}^*$ by the convergent Taylor series in the vicinity of $\underline{s}$,

(5.5.1.6) $$\underset{\sim}{r}^* = \underset{\sim}{r}(X^1, X^2; t) + \sum_{N=1}^{\infty} \frac{1}{N!} X^N \left(\frac{\partial^N \underset{\sim}{r}^*}{\partial X^N} \right)_{X=0} .$$

The quantities

(5.5.1.7) $$\frac{1}{N!} \left(\frac{\partial^N \underset{\sim}{r}^*}{\partial X^N} \right)_{X=0} = \underset{\sim}{d}_{(\underset{\sim}{N})}$$

may be called directors, and we see that they are functions of coordinates of the points on the middle surface $\underline{s}$,

(5.5.1.8) $$\underset{\sim}{d}_{(N)} = \underset{\sim}{d}_{(N)}(X^1, X^2; t) .$$

At any instant of time t the configuration of a shell is completely determined by the configuration of the surface $\underline{s}$ and by the directors $\underset{\sim}{d}_{(N)}$.

The velocity vector at a point $\underset{\sim}{X}$ of the shell will, according to (5.5.1.6) be

(5.5.1.9) $$\underset{\sim}{v}^* = \dot{\underset{\sim}{r}}^* = \underset{\sim}{v} + \sum_{N=1}^{\infty} X^N \dot{\underset{\sim}{d}}_{(N)} ,$$

where

(5.5.1.10) $$\underset{\sim}{v} = \dot{\underset{\sim}{r}}(X^1, X^2; t) .$$

We shall define the base vectors $\underset{\sim}{g}_k$ at the points of the shell by

$$\underset{\sim}{g}_k = \frac{\partial \underset{\sim}{r}^*}{\partial x^k} , \qquad (5.5.1.11)$$

and the base vectors $\underset{\sim}{a}_\alpha$ at the points of the surface $\underline{s}$ by

$$\underset{\sim}{a}_\alpha = \frac{\partial \underset{\sim}{r}}{\partial x^\alpha} . \qquad (5.5.1.12)$$

It follows from (5.5.1.6) that

$$\underset{\sim}{g}_\alpha = \underset{\sim}{a}_\alpha + \sum_{N=1}^{\infty} x^N \frac{\partial \underset{\sim}{d}_{(N)}}{\partial x^\alpha} , \qquad (5.5.1.13)$$

$$\underset{\sim}{g}_3 = \sum_{N=1}^{\infty} N x^{N-1} \underset{\sim}{d}_{(N)} .$$

Let $\varrho^*(\underset{\sim}{r}^*)$ be the density of matter at the points of the shell. The momentum of any part v of the shell, bounded by the surfaces $\alpha \leq x \leq \beta$ and by a contour $\underline{c}$ enclosing an area σ of the surface $\underline{s}$ will be

$$\underset{\sim}{K} = \int_v \varrho^* \underset{\sim}{v}^* dv = \int_\sigma\int_\alpha^\beta \varrho^* \sqrt{g}\Big(\underset{\sim}{v} + \sum_{N=1}^{\infty} X^N \dot{\underset{\sim}{d}}_{(N)}\Big) dX^1 dX^2 dX . \quad (5.5.1.14)$$

The vectors $\underset{\sim}{v}$ and $\dot{\underset{\sim}{d}}_{(N)}$ are independent of X, and we may put

$$\int_\alpha^\beta \varrho^* \sqrt{g}\, dX = \varrho \sqrt{a} , \qquad (5.5.1.15)$$

where ϱ is the density of matter per unit area of the surface $\underline{s}$ and

$$g = \det g_{ij} , \quad g_{ij} = \underset{\sim}{g}_i \cdot \underset{\sim}{g}_j , \qquad (5.5.1.16a)$$

(5.5.1.16b) $$a = \det a_{\alpha\beta}\ ,\quad a_{\alpha\beta} = \underset{\sim}{a}_\alpha \cdot \underset{\sim}{a}_\beta\ .$$

The quantities ϱ and a are functions of X^1 and X^2 only. We shall also write

(5.5.1.17) $$\int_\alpha^\beta \varrho^* X^N \sqrt{g}\, dX = \varrho k^N \sqrt{a}\ ,\quad (N = 2,3,4,\ldots)$$

where

(5.5.1.18) $$\int_\alpha^\beta \varrho^* X \sqrt{g}\, dX = 0\ .$$

The last relation fixes the surface $\underline{s}$ with respect to the bounding surfaces α and β. The quantities k^N are functions of X^1 and X^2 only.

From (5.5.1.14) we have now

(5.5.1.19) $$\underset{\sim}{K} = \int_\sigma \varrho\Big(\underset{\sim}{v} + \sum_{N=2}^{\infty} k^N \dot{\underset{\sim}{d}}_{(N)}\Big) d\sigma\ .$$

For the moment of momentum we have now

$$\underset{\sim}{l}^0 = \int_v \varrho^* \underset{\sim}{r}^* \times \underset{\sim}{v}^* dv =$$

(5.5.1.20) $$= \int_\sigma \int_\alpha^\beta \varrho^* \sqrt{g} \Big[\underset{\sim}{r} \times \underset{\sim}{v} + \sum_{M=1}^{\infty} X^M (\underset{\sim}{r} \times \dot{\underset{\sim}{d}}_{(M)} + \underset{\sim}{d}_{(M)} \times \underset{\sim}{v}) + \sum_{M,N=1}^{\infty} X^{N+M} \underset{\sim}{d}_{(N)} \times \dot{\underset{\sim}{d}}_{(M)}\Big] dX^1 dX^2 dX\ ,$$

and if we introduce the notation

$$\int_{\alpha}^{\beta} \varrho^{*} X^{N+M} \sqrt{g}\, dX = \varrho k^{N+M} \sqrt{a}\,, \qquad (5.5.1.21)$$

for the moment of momentum we obtain the expression

$$\underset{\sim}{\ell}^{0} = \int_{\sigma} \varrho \Big[\underset{\sim}{r} \times \underset{\sim}{v} + \sum_{N=2}^{\infty} k^{N} \big(\underset{\sim}{r} \times \dot{\underset{\sim}{d}}_{(N)} + \underset{\sim}{d}_{(N)} \times \underset{\sim}{v} \big) + \sum_{M,N=1}^{\infty} k^{N+M} \underset{\sim}{d}_{(N)} \times \dot{\underset{\sim}{d}}_{(M)} \Big] d\sigma\,. \qquad (5.5.1.22)$$

Using the same notation and procedure, we find for the kinetic energy of the considered portion of the shell the following expression

$$T = \frac{1}{2} \int_{\sigma} \varrho \Big(\underset{\sim}{v} \cdot \underset{\sim}{v} + 2 \sum_{N=2}^{\infty} k^{N} \underset{\sim}{v} \cdot \dot{\underset{\sim}{d}}_{(N)} + \sum_{N,M=1}^{\infty} k^{N+M} \dot{\underset{\sim}{d}}_{(N)} \cdot \dot{\underset{\sim}{d}}_{(M)} \Big) d\sigma\,. \quad (5.5.1.23)$$

If $\underset{\sim}{D}_{(1)}$, $\underset{\sim}{D}_{(2)}$, ... are directors in the initial configuration, and if

$$\underset{\sim}{R}^{*} = \underset{\sim}{R} + \sum_{N=1}^{\infty} X^{N} \underset{\sim}{D}_{(N)}$$

is the position vector for points of the shell in the initial configuration, the equations of motion may be considered in the form

$$\underset{\sim}{r} = \underset{\sim}{r}(\underset{\sim}{R}, t)\,; \qquad \underset{\sim}{d}_{(N)} = \underset{\sim}{d}_{(N)} \big[\underset{\sim}{D}_{(N)}(X^{\alpha}); t \big]\,. \qquad (5.5.1.24)$$

Retaining in (5.5.1.6) only the terms linear in X we see that in this approximation all points of the shell which were in the initial configuration situated on the straight line $\underset{\sim}{D}_{(1)} X$, in the deformed configuration will be again on the straight line $\underset{\sim}{d}_{(1)} X$. The higher approximation in (5.5.1.6) we take, the more precise description of the distribution of the points of the shell outside the middle surface $\underline{s}$ we obtain. In the linear approximation the expressions (5.5.1.19, 22, 23) will obtain the form analogous to (5.1.24, 26), or to (5.2.23), but for a medium with a single director field.

Some other contemporary approaches to the shell theory, such as Reissner's (see Section 12), which is partly based on the earlier work of Günther [189, 190] and Schäfer [390] may be considered as a special case of the here outlined general approach. Reissner regards shells as Cosserat bodies with rigid director triads. In that case the configuration of a shell is described in terms of the position vector $\underset{\sim}{r}$ of points on the middle surface, and in terms of the rotation vector $\underset{\sim}{\Phi}$, which is independent of the displacements of points on s and describes the rotations of shell elements (cf. Reissner [368, 370, 371], Reissner and Wan [375, 376], Wan [479, 480, 481]).

5.5.2 Rods

The basic ideas for the theory of rods are essentially the same as for the general theory of shells, sketched above. Let

(5.5.2.1) $$X^{\alpha} = 0 , \quad \alpha = 1,2$$

be the parametric equations of a smooth and non-intersecting curve C in the space; we consider this curve as the middle curve of a rod. The position vector of any point of the rod in the initial configuration is

(5.5.2.2) $$\underset{\sim}{R}^{*} = \underset{\sim}{R}^{*}(X^{1},X^{2},X) ,$$

where $X = X^3$ is the parameter varying along C. It is assumed that $\underset{\sim}{R}^*$ for sufficiently small values of X^α may be expanded into a series

$$\underset{\sim}{R}^* = \underset{\sim}{R}(0,0,X) + X^\alpha \frac{\partial \underset{\sim}{R}^*}{\partial X^\alpha} + \frac{1}{2} X^{\alpha_1} X^{\alpha_2} \frac{\partial^2 \underset{\sim}{R}^*}{\partial X^{\alpha_1} \partial X^{\alpha_2}} + \ldots, \quad (5.5.2.3)$$

where $\underset{\sim}{R}$ is the position vector of any point on C.

Introducing the notation

$$\frac{1}{n!}\left(\frac{\partial^n \underset{\sim}{R}^*}{\partial X^{\alpha_1} \ldots \partial X^{\alpha_n}}\right)_{X^\alpha = 0} = \underset{\sim}{D}_{\alpha_1 \ldots \alpha_n} \quad (5.5.2.4)$$

we may write

$$\underset{\sim}{R}^\alpha = \underset{\sim}{R} + \sum_{n=1}^{\infty} X^{\alpha_1} \ldots X^{\alpha_n} \underset{\sim}{D}_{\alpha_1 \ldots \alpha_n} . \quad (5.5.2.5)$$

At a time t the curve C will be c, and the position vector of points of c will be $\underset{\sim}{r}$, such that

$$\underset{\sim}{r}^* = \underset{\sim}{r}^*(X^1, X^2, X; t) = \underset{\sim}{r}(0,0,X;t) + \sum_{n=1}^{\infty} X^{\alpha_1} \ldots X^{\alpha_n} \underset{\sim}{d}_{\alpha_1 \ldots \alpha_n}, \quad (5.5.2.6)$$

where we have put

$$\underset{\sim}{d}_{\alpha_1 \ldots \alpha_n} = \frac{1}{n!}\left(\frac{\partial^n \underset{\sim}{r}^*}{\partial X^{\alpha_1} \ldots \partial X^{\alpha_n}}\right)_{X^\alpha = 0} . \quad (5.5.2.7)$$

The directors $\underset{\sim}{d}_{\alpha_1 \ldots \alpha_n}$ are functions of the variable X along c and of the time t. Here again $x^k = \delta^k_K X^K$ are considered coordinates.

The base vectors at the points of the rod are

$$\text{(5.5.2.8)} \qquad \underset{\sim}{g}_k = \frac{\partial \underset{\sim}{r}^*}{\partial x^k} ,$$

and the tangential vector $\underset{\sim}{a}$ to the middle curve $\underline{c}$ is given by

$$\text{(5.5.2.9)} \qquad \underset{\sim}{a} = \underset{\sim}{a}_3 = \frac{\partial \underset{\sim}{r}}{\partial X} , \quad (a_{33} = \underset{\sim}{a}_3 \cdot \underset{\sim}{a}_3) .$$

From the last two relations we find

$$\text{(5.5.2.10)} \qquad \begin{aligned} \underset{\sim}{g}_\beta &= \underset{\sim}{a}_\beta + \sum_{n=2}^{\infty} n X^{\alpha_2} \dots X^{\alpha_n} \underset{\sim}{d}_{\beta\alpha_2 \dots \alpha_n} , \\ \underset{\sim}{g}_3 &= \underset{\sim}{a} + \sum_{n=1}^{\infty} X^{\alpha_1} \dots X^{\alpha_n} \frac{\partial \underset{\sim}{d}_{\alpha_1 \dots \alpha_n}}{\partial X} . \end{aligned}$$

We assume that the rod is a three-dimensional body bounded by a surface

$$\text{(5.5.2.11)} \qquad f(X^1, X^2) = 0$$

such that X =const. represents curved sections σ bounded by closed curves. We shall consider an arbitrary element of the rod bounded by $\alpha \leq X \leq \beta$ and by the surface (5.5.2.11).

The momentum of the considered element of the rod will be the vector

$$\text{(5.5.2.12)} \quad \underset{\sim}{K} = \int_v \varrho^* \underset{\sim}{v}^* dv = \int_v \varrho^* \sqrt{g} \Big(\underset{\sim}{v} + \sum_n X^{\alpha_1} \dots X^{\alpha_n} \dot{\underset{\sim}{d}}_{\alpha_1 \dots \alpha_n} \Big) dv .$$

Since $\underset{\sim}{v}$ and $\underset{\sim}{d}_{\alpha_1 \dots \alpha_n}$ are independent of X^1, X^2, we may write

$$\text{(5.5.2.13)} \qquad \iint_\sigma \varrho^* \sqrt{g}\, dX^1 dX^2 = \varrho \sqrt{a_{33}} ,$$

$$\iint_{\sigma} \varrho^* \sqrt{g}\, X^{\alpha_1} \dots X^{\alpha_n} dX^1 dX^2 = \varrho k^{\alpha_1 \dots \alpha_n} \sqrt{a_{33}} \quad , \quad (5.5.2.14)$$

and

$$\underset{\sim}{K} = \int_{\alpha}^{\beta} \varrho \left(\underset{\sim}{v} + \sum_{n=1}^{\infty} k^{\alpha_1 \dots \alpha_n} \dot{\underset{\sim}{d}}_{\alpha_1 \dots \alpha_n} \right) \sqrt{a_{33}}\, dX \ . \qquad (5.5.2.15)$$

The expressions for the moment of momentum will be obtained from

$$\begin{aligned} \underset{\sim}{\ell}^0 &= \int_{v} \varrho^* \underset{\sim}{r}^* \times \underset{\sim}{v}^* dv \\ &= \int_{v} \varrho^* \left(\underset{\sim}{r} + \sum_n X^{\alpha_1} \dots X^{\alpha_n} \underset{\sim}{d}_{\alpha_1 \dots \alpha_n} \right) \times \left(\underset{\sim}{v} + \sum_m X^{\beta_1} \dots X^{\beta_m} \dot{d}_{\beta_1 \dots \beta_m} \right) dv \ , \end{aligned} \qquad (5.5.2.16)$$

and using (5.5.2.14) we may write it in the form

$$\begin{aligned} \underset{\sim}{\ell}^0 = \int_{\alpha}^{\beta} \varrho \Big(& \underset{\sim}{t} \times \underset{\sim}{v} + \sum_n k^{\alpha_1 \dots \alpha_n} \underset{\sim}{r} \times \dot{\underset{\sim}{d}}_{\alpha_1 \dots \alpha_n} + \sum_n k^{\alpha_1 \dots \alpha_n} \underset{\sim}{d}_{\alpha_1 \dots \alpha_n} \times \underset{\sim}{v} + \\ & + \sum_{n,m} k^{\alpha_1 \dots \alpha_n \beta_1 \dots \beta_m} \underset{\sim}{d}_{\alpha_1 \dots \alpha_n} \times \dot{\underset{\sim}{d}}_{\beta_1 \dots \beta_m} \Big) \sqrt{a_{33}}\, dX \ . \end{aligned} \qquad (5.5.2.17)$$

Applying the same procedure, for the kinetic energy of the considered section the rod we find the expression

$$\begin{aligned} T = \frac{1}{2} \int_{\alpha}^{\beta} \varrho \Big(& \underset{\sim}{v} \cdot \underset{\sim}{v} + 2 \sum_{n=2}^{\infty} k^{\alpha_1 \dots \alpha_n} \underset{\sim}{v} \cdot \dot{\underset{\sim}{d}}_{\alpha_1 \dots \alpha_n} + \\ & + \sum_{n,m=1}^{\infty} k^{\alpha_1 \dots \alpha_n \beta_1 \dots \beta_m} \dot{\underset{\sim}{d}}_{\alpha_1 \dots \alpha_n} \cdot \dot{\underset{\sim}{d}}_{\beta_1 \dots \beta_m} \Big) \sqrt{a_{33}}\, dX \ . \end{aligned} \qquad (5.5.2.18)$$

The linear approximation with respect to X^{α} leads to the representation of rods in which we consider instead of rods curves with two directors attached to their points (cf.

Section 7.3).

6. Forces Stresses and Couples

In the mechanics of particles it is usually proved that a system of forces, say $\underset{\sim}{f}_{(1)}, \underset{\sim}{f}_{(2)}, \dots, \underset{\sim}{f}_{(n)}$ acting on a system of particles $M_1, \dots, M_n$ may be reduced to the resultant force

(6.1) $$\underset{\sim}{f} = \sum_{i=1}^{n} \underset{\sim}{f}_{(i)}$$

and to the resultant couple, which is defined with respect to a pole $\underset{\sim}{O}$ by the expression

(6.2) $$\underset{\sim}{M}^{0} = \sum_{i=1}^{n} \underset{\sim}{r}_i \times \underset{\sim}{f}_{(i)} ,$$

where $\underset{\sim}{r}_i$ are position vectors of the particles M_i with respect to $\underset{\sim}{O}$. In continuum mechanics an immediate generalization is insufficient to describe all the forces and couples which appear, even if the suitable assumptions are made for the transition from a discrete system to a continuum model.

In the following definition we partly follow Truesdell and Noll [379], but we introduce some additional definitions in order to consider more general models of continua.

Let v be a part of a body B and S the bounding surface of the v, and let the motion of the body be given by the equations

(6.3a) $$x^i = x^i(\underset{\sim}{X}, t) ,$$

$$d^i_{(\alpha)} = d^i_{(\alpha)}(\underset{\sim}{X},t) , \quad (\alpha = 1,2,..n) \tag{6.3b}$$

and let $\varrho = \varrho_{(\underset{\sim}{x})}$ be the density of matter.

1. At each time t there is a vector field $\underset{\sim}{f}(\underset{\sim}{x},t)$ defined per unit mass, which we call the <u>external body force</u>. The vector $\underset{\sim}{F}_F(v)$ defined by the volume integral

$$\underset{\sim}{F}_F(v) = \int_v \varrho \underset{\sim}{f}(\underset{\sim}{x})dv \tag{6.4}$$

is called the <u>resultant external body force</u> exerted on the part v at time t.

2. At each time t there is an antisymmetric tensor field $l^{ij}(\underset{\sim}{x},t)$ defined per unit mass, which we call the <u>external body couple</u>. The <u>resultant body couple</u> is defined by the volume integral

$$M_l^{ij}(v) = \int_v \varrho l^{ij}(\underset{\sim}{x})dv . \tag{6.5}$$

3. At each time t, to each part v of the body B corresponds a vector field $\underset{\sim}{t}(x,t)$, defined for the points $\underset{\sim}{x}$ on the bounding surface s of v. It is called the <u>stress</u> (or the <u>density of the contact force</u>), acting on the part v of B. The <u>resultant contact force</u> $F_t(v)$ exerted on v at time t is defined by the surface integral

$$F_t(v) = \oint_s \underset{\sim}{t}(\underset{\sim}{x},v)ds . \tag{6.6}$$

4. At each time t, to each part v of the body B corresponds an antisymmetric tensor field m^{ij} defined for the point $\underset{\sim}{x}$ on the boundary s of v. It is called the couple stress (or the density of the contact couple) acting on the part v of B. The resultant contact couple $M_m^{ij}(v)$ is defined by the surface integral

(6.7) $$M_m^{ij}(v) = \oint_s m^{ij}(\underset{\sim}{x}, v)\,ds \, .$$

5. The total resultant force exerted on the part v of B is defined as the sum of the resultant body force and the resultant contact force,

(6.8) $$\underset{\sim}{F}(v) = \underset{\sim}{F}_f(v) + \underset{\sim}{F}_t(v) \, .$$

6. The total resultant couple exerted on the part v of B is defined as the sum of the resultant body couple and the resultant contact couple,

(6.9) $$M^{ij}(v) = M_\ell^{ij}(v) + M_m^{ij(v)} .$$

According to the stress principle (cf. [469] there is a vector field $\underset{\sim}{t}(\underset{\sim}{x}, \underset{\sim}{n})$ defined for all points $\underset{\sim}{x}$ in B and for all unit vectors $\underset{\sim}{n}$ such that the stress acting on any part v of B is given by

(6.10) $$\underset{\sim}{t}(\underset{\sim}{x}, v) = \underset{\sim}{t}(\underset{\sim}{x}, n) \, ,$$

where $\underset{\sim}{n}$ is the exterior unit normal vector at the points $\underset{\sim}{x}$ on the boundary of S.

In elementary continuum mechanics it is proved that the stress vector $\underset{\sim}{t}(\underset{\sim}{x},\underset{\sim}{n})$,

$$\underset{\sim}{t}(\underset{\sim}{x},\underset{\sim}{n}) = t^{i}(\underset{\sim}{x},\underset{\sim}{n})\underset{\sim}{g}_{i} \tag{6.11}$$

may be represented in the form

$$\underset{\sim}{t}(\underset{\sim}{x},\underset{\sim}{n}) = t^{ij}(\underset{\sim}{x})n_{j}\underset{\sim}{g}_{i} \;, \tag{6.12}$$

where $t^{ij}(\underset{\sim}{x})$ are components of the stress tensor. From (6.6) we obtain now that the components of the resultant stress are given by the integral

$$F_{t}(v) = \oint_{S} t^{ij}(\underset{\sim}{x})\underset{\sim}{g}_{i}n_{j}ds \;. \tag{6.13}$$

In analogy to the stress vector we may write for the couple stress

$$m^{ij}(\underset{\sim}{x},v) = m^{ij}(\underset{\sim}{x},\underset{\sim}{n}) \;, \tag{6.14}$$

and

$$m^{ij}(\underset{\sim}{x},\underset{\sim}{n}) = m^{ijk}(\underset{\sim}{x})n_{k} \tag{6.15}$$

where $m^{ijk} = -m^{jik}$ is the couple-stress tensor (cf. [469]).

7. At each time t, at each part v of the body B there are vector fields $\underset{\sim}{K}^{(\alpha)}(\underset{\sim}{x},t)$ defined per unit mass, which we call the external director forces. The vectors $\underset{\sim}{F}^{\alpha}_{k}(v)$ defined by

the integral

$$F^{\alpha}_{k}(v) = \int_{v} \varrho \underset{\sim}{k}^{(\alpha)}(\underset{\sim}{x}, t)\, dv , \quad (\alpha = 1, 2, \ldots, n) \tag{6.16}$$

are called the resultant director forces exerted on the part v of the body at time t.

8. At each time t, to each part v of the body B correspond vector fields $\underset{\sim}{h}^{(\alpha)}(\underset{\sim}{x}, v)$, defined for the points $\underset{\sim}{x}$ on the boundary s of v, which we call the director stresses. We assume that there are vector fields $\underset{\sim}{h}^{(\alpha)}(\underset{\sim}{x}, v)$, defined for all points of v and for all unit vectors $\underset{\sim}{n}$, such that the director stresses acting on any part v of B are given by

$$\underset{\sim}{h}^{(\alpha)}(\underset{\sim}{x}, v) = \underset{\sim}{h}^{(\alpha)}(\underset{\sim}{x}, \underset{\sim}{n}) , \quad (\alpha = 1, 2, \ldots, n) . \tag{6.17}$$

The resultant director stresses are given by the surface integrals

$$F^{\alpha}_{h}(v) = \oint_{s} \underset{\sim}{h}^{(\alpha)}(\underset{\sim}{x}, v)\, ds , \quad (\alpha = 1, 2, \ldots, n) . \tag{6.18}$$

For the director stress vectors $\underset{\sim}{h}^{(\alpha)}(\underset{\sim}{x}, \underset{\sim}{n})$ we assume that they may be represented in the form

$$\underset{\sim}{h}^{(\alpha)}(\underset{\sim}{x}, \underset{\sim}{n}) = h^{(\alpha)i}(\underset{\sim}{x}, \underset{\sim}{n})\underset{\sim}{g}_i , \tag{6.19}$$

and that

$$\underset{\sim}{h}^{(\alpha)}(\underset{\sim}{x}, v) = \underset{\sim}{h}^{(\alpha)}(\underset{\sim}{x}, \underset{\sim}{n}) = h^{(\alpha)ij}(\underset{\sim}{x})\underset{\sim}{g}_i n_j . \tag{6.20}$$

The quantities $h^{(\alpha)ij}$ we call the director stress tensors.

9. The total resultant director forces exerted on the part v of B are defined as the sum of the resultant director forces and the resultant director stresses,

$$F^{\alpha}_{\underset{\sim}{d}}(v) = F^{\alpha}_{k}(v) + F^{\alpha}_{h}(v) . \tag{6.21}$$

We assume that the number of the director force vectors and of the director stress tensors is equal to the number of the directors $\underset{\sim}{d}_{(\alpha)}$ of the body .

The momenta of forces and stresses are defined by the following expressions:

a) The moment of the external body force at a point $\underset{\sim}{x}$, with respect to the origin $\underset{\sim}{0}$:

$$\underset{\sim}{r} \times \varrho \underset{\sim}{f} , \tag{6.22}$$

and the resultant moment for the part v of B

$$\int_v \varrho \underset{\sim}{r} \times \underset{\sim}{f} \, dv . \tag{6.23}$$

b) The moment of stress at $\underset{\sim}{x}$, with respect to the origin $\underset{\sim}{0}$:

$$\underset{\sim}{r} \times \underset{\sim}{t}(\underset{\sim}{x}, \underset{\sim}{n}) , \tag{6.24}$$

and the resultant moment of stress :

$$\oint_S \underset{\sim}{r} \times \underset{\sim}{t}(\underset{\sim}{x}, \underset{\sim}{n}) \, ds . \tag{6.25}$$

c) The moment of the director forces at $\mathbf{x}$:

$$\varrho\boldsymbol{\Gamma} = \mathbf{d}_{(\alpha)}\times\varrho\mathbf{k}^{(\alpha)}(\mathbf{x},t) \tag{6.26}$$

and the resultant of the director forces for the part v of B:

$$\int_v \varrho\,\mathbf{d}_{(\alpha)}\times\mathbf{k}^{(\alpha)}(\mathbf{x},t)\,dv\ . \tag{6.27}$$

d) The moment of the director stresses at $\mathbf{x}$:

$$\mathbf{d}_{(\alpha)}\times\mathbf{h}^{(\alpha)}(\mathbf{x},\mathbf{n})\ , \tag{6.28}$$

and the resultant moment of the director stresses,

$$\oint_S \mathbf{d}_{(\alpha)}\times\mathbf{h}^{(\alpha)}(\mathbf{x},\mathbf{n})\,ds\ . \tag{6.29}$$

The total resultant moment of forces acting on a part v of a body B at time t is the sum of the moments of body and director forces, of body and director couples, and of the moments of stress and director stresses, and of the couple stresses,

$$\begin{aligned}\mathbf{L} = {} & \int_v \varrho(\mathbf{r}\times\mathbf{f} + \mathbf{d}_{(\lambda)}\times\mathbf{k}^{(\lambda)} + \boldsymbol{\ell})\,dv\ + \\ & + \oint_S (\mathbf{r}\times\mathbf{t} + \mathbf{d}_{(\alpha)}\times\mathbf{h}^{(\alpha)} + \mathbf{m})\,ds\ .\end{aligned} \tag{6.30}$$

This may be written in the component form as follows:

$$\begin{aligned}L^{\alpha\beta} = {} & 2\int_v \varrho(z^{[\alpha}f^{\beta]} + d^{[\alpha}_{(\lambda)}k^{(\lambda)\beta]} + \ell^{\alpha\beta})\,dv\ + \\ & + 2\oint_S (z^{[\alpha}t^{\beta]\gamma} + d^{[\alpha}_{(\lambda)}h^{(\lambda)\beta]\gamma} + m^{\alpha\beta\gamma})n_\gamma\,ds\ .\end{aligned}$$

6.1 A Physical Interpretation

Physical interpretations of the director forces depend on the model considered. For a medium consisting of particles which are composed of mass points, as was the medium considered in the section 5.1, we may assume (Rivlin [377, 378]) that the external force $m_\alpha^{(P)} \underset{\sim}{f}_\alpha^{(P)}$ acts on the mass point $m_\alpha^{(P)}$ of the Pth particle. The resultant external force acting on the Pth particle is

$$\sum_{\alpha=1}^{n} m_\alpha^{(P)} \underset{\sim}{f}_\alpha^{(P)} = m^{(P)} \underset{\sim}{f}^{(P)} , \tag{6.1.1}$$

and if we assume that the discrete sets of vectors $\underset{\sim}{f}^{(P)}$ and $\underset{\sim}{f}_\alpha^{(P)}$ may be replaced by continuous vector fields $\underset{\sim}{f}$ and $\underset{\sim}{f}_\alpha$, defined throughout the body B , for a part v of B we may write for the resultant body force

$$\underset{\sim}{F}_f(v) = \sum_v m^{(P)} \underset{\sim}{f}^{(P)} = \int_v \varrho \underset{\sim}{f} \, dv . \tag{6.1.2}$$

Denoting again by $\underset{\sim}{r}^{(P)}$ the position vectors of the centres of mass of the particle and by $\underset{\sim}{\varrho}_\alpha^{(P)}$ the position vectors of the mass points inside the particles, with respect to the corresponding centres of mass, the moment of the force $m_\alpha^{(P)} \underset{\sim}{f}_\alpha^{(P)}$ with respect to the origin $\underset{\sim}{O}$ will be

$$(\underset{\sim}{r}^{(P)} + \underset{\sim}{\varrho}_\alpha^{(P)}) \times m_\alpha^{(P)} \underset{\sim}{f}_\alpha^{(P)} . \tag{6.1.3}$$

For a particle P we have for the resultant moment

of external forces the expression

$$(6.1.4) \qquad \underset{\sim}{r}^{(P)} \times m^{(P)} \underset{\sim}{f}^{(P)} + \sum_{\alpha=1}^{n} \underset{\sim}{\varrho}_{\alpha}^{(P)} \times m_{\alpha}^{(P)} \underset{\sim}{f}_{\alpha}^{(P)} ,$$

and for the part v of B under the suitable assumptions we may write

$$(6.1.5) \qquad \sum_{v} (\underset{\sim}{r}^{(P)} \times m^{(P)} \underset{\sim}{f}^{(P)} + \sum_{\alpha=1}^{n} \underset{\sim}{\varrho}_{\alpha}^{(P)} \times m_{\alpha}^{(P)} \underset{\sim}{f}_{\alpha}^{(P)}) =$$
$$= \int_{v} \varrho \underset{\sim}{r} \times \underset{\sim}{f}\, dv + \int_{v} \varrho \underset{\sim}{d}_{(\alpha)} \times f_{\alpha}\, dv ,$$

where according to the section 5.1 the discrete vectors $\underset{\sim}{\varrho}_{\alpha}^{(P)}$ are replaced by continuous vector fields $\underset{\sim}{d}_{(\alpha)}$.

According to Rivlin [378], the field $\underset{\sim}{f}$ represents the body force field, and $\underset{\sim}{f}_{\alpha}$ are the director force fields.

According to this model of Rivlin's, if s is the bounding surface of v in B, under the assumption that <u>on the surface</u> s the discrete vectors $\underset{\sim}{f}^{(P)}$ and $\underset{\sim}{f}_{\alpha}^{(P)}$ may be replaced by continuous vector fields $\underset{\sim}{t}$ and $\underset{\sim}{t}_{(\alpha)}$, we may write

$$(6.1.6) \qquad \sum_{S} m^{(P)} \underset{\sim}{f}^{(P)} = \oint_{S} \underset{\sim}{t} \cdot ds ,$$

where $d\underset{\sim}{s} = \underset{\sim}{n} ds$ is the directed surface element and $\underset{\sim}{n}$ is the unit vector, and

$$(6.1.7) \qquad \sum_{S} m_{\alpha}^{(P)} \underset{\sim}{\varrho}_{\alpha}^{(P)} \times \underset{\sim}{f}_{\alpha}^{(P)} = \oint_{S} \underset{\sim}{d}_{(\alpha)} \times \underset{\sim}{t}_{(\alpha)} ds .$$

$\underset{\sim}{t}$ represents the <u>simple surface force field</u>, or the stress, and $\underset{\sim}{t}_{(\alpha)}$ are the <u>director surface force fields</u>, or the director stresses according to the terminology introduced in the previous section.

7. Balance and Conservation Principles

The differential equations of motion in classical continuum mechanics are usually derived from the law of conservation of mass (equation of continuity), and from the Euler's laws of balance of momentum and moment of momentum. Since we postulate here the validity of these laws, we regard them as principles.

Let v be a part of a body B and s the boundary of v. Let $\underset{\sim}{T}$ be the density of a quantity in balance, $\underset{\sim}{A}$ its influx (or efflux) per unit area of the bounding surface and $\underset{\sim}{B}$ its source per unit volume. The equation of balance has the general form

$$\frac{d}{dt}\int_v \underset{\sim}{T}\,dv = \oint_s \underset{\sim}{A}\cdot d\underset{\sim}{s} + \int_v \underset{\sim}{B}\,dv \tag{7.1}$$

where $d\underset{\sim}{s}$ is the oriented surface element, $d\underset{\sim}{s} = \underset{\sim}{n}\,ds$, and $\underset{\sim}{n}$ the unit normal vector to ds. If the source vanishes, the equation of balance becomes the equation of conservation.

In classical mechanics we assume that there are neither sources nor influxes of mass. If ϱ is the density of mass, so that dm,

$$\varrho\,dv = dm\,, \tag{7.2}$$

is the mass contained in the volume dv, the mass contained in the part v of the body considered will be

$$m(v) = \int_v \varrho dv . \tag{7.3}$$

From (7.1) we may write now the law of conservation of mass,

$$\frac{dm}{dt} = \frac{d}{dt}\int_v \varrho dv = 0 ,$$

which may be written in the form

$$\int_v (\dot{\varrho} dv + \varrho \dot{\overline{dv}}) = 0 . \tag{7.4}$$

For a body in motion the equations of motion of its points are

$$x^i = x^i(X^1,X^2,X^3,t) , \quad (i = 1,2,3) \tag{7.5}$$

where X^K are material, and x^k spatial coordinates. If dV is the volume element of the body in an initial configuration referred to the coordinates X^k, and dv the corresponding volume element in a configuration $K(t)$ at time t, the volume elements dv and dV are related by the formula

$$dv = JdV , \tag{7.6}$$

where

$$J = \sqrt{\frac{g}{G}}\det(x^k{}_{;K}) . \tag{7.7}$$

From (7.6) we have now

$$\dot{\overline{dv}} = \dot{J} dV , \tag{7.8}$$

and since*

$$\dot{J} = J \dot{x}^k_{;K} = J \operatorname{div} \underset{\sim}{v} , \tag{7.9}$$

from (7.4) we immediately have the global form of the law of conservation of mass

$$\int_v (\dot{\varrho} + \varrho v^k_{;k}) dv = 0 \tag{7.10}$$

this has to be valid for an arbitrary part v of the body and therefore we finally obtain the local form of this law, which is often called the <u>equation of continuity</u>,

$$\dot{\varrho} + \varrho v^k_{;k} = 0 . \tag{7.11}$$

In general, the density ϱ is a function of position and time, $\varrho = \varrho(\underset{\sim}{x}, t)$ and $\dot{\varrho} = \partial\varrho/\partial t + \varrho_{,k} v^k$. Substituting this in (7.11) we obtain the continuity equation in another

---

*According to the rule for the differentiation of determinants, if $a = \det a^i_{;j}$, then $\dot{a}\delta^i_j = \dot{a}^i_j A^j_{;k}$, where $A^j_{;k}$ is the cofactor in a corresponding to the element $a^k_{;j}$. Since $X^K_{;k}$ = (cofactor for $x^k_{;K}$) / (det $x^k_{;K}$), we have

$$\dot{\overline{\det x^k_{;K}}} = \dot{x}^k_{;K} X^K_{;\ell} (\det x^m_{;M}) \delta^\ell_k = v^k_{;k} \det x^m_{;M}$$

where $v^k = \dot{x}^k$ is the velocity vector.

form,

$$\frac{\partial \varrho}{\partial t} + (\varrho v^k)_{,k} = 0 . \tag{7.12}$$

The <u>principle of balance of momentum</u> states that the rate of the global momentum $\underset{\sim}{K}$ of a part v of a body B is equal to the total resultant force exerted on the part v of the body. According to (6.4), (6.6), (6.8) and (6.10), for the total resultant force we have

$$\underset{\sim}{F}(v) = \int_v \varrho \underset{\sim}{f} dv + \oint_S \underset{\sim}{t}(\underset{\sim}{x}, \underset{\sim}{n}) ds . \tag{7.13}$$

We assume the momentum $\underset{\sim}{K}$ of a part v of a body B to have the form given by (5.1.24) or (5.2.23)

$$\underset{\sim}{K} = \int_v \varrho \underset{\sim}{v} dv ,$$

and the balance of momentum equation reads

$$\frac{d}{dt}\int_v \varrho \underset{\sim}{v} dv = \int_v \varrho \underset{\sim}{f} dv + \oint_S \underset{\sim}{t}(\underset{\sim}{x}, \underset{\sim}{n}) ds . \tag{7.14}$$

Using (6.12) and referring for the sake of simplicity all quantities to a Cartesian system of reference z^α; the component form of (7.14) becomes

$$\frac{d}{dt}\int_v \varrho \dot{z}^\alpha dv = \int_v \varrho f^\alpha dv + \oint_S t^{\alpha\beta} n_\beta ds . \tag{7.15}$$

Performing the differentiation on the left-hand side and applying the divergence theorem to the surface integral on the right--hand side of (7.16), and using the continuity equation (7.11) we obtain

$$\int_v \varrho \dot{v}^\alpha dv = \int_v (\varrho f^\alpha + t^{\alpha\beta}{}_{,\beta}) dv , \tag{7.17}$$

which is valid for an arbitrary part v of B and therefore the relation (7.17) must be valid at all points of B, which give the local equation for the balance of momentum;

$$\varrho \dot{v}^\alpha = t^{\alpha\beta}{}_{,\beta} + \varrho f^\alpha . \tag{7.18}$$

This is a tensorial equation and for arbitrary curvilinear coordinates x^i we have

$$\varrho \dot{v}^i = t^{ij}{}_{;j} + \varrho f^i , \tag{7.19}$$

where (see Appendix, (A3.10))

$$\dot{v}^i = \frac{\partial v^i}{\partial t} + v^i{}_{;j} v^j , \tag{7.20}$$

and $t^{ij}{}_{;j}$ represents the covariant derivative of $\underset{\sim}{t}$ with respect to x^j, or the divergence of the tensor $\underset{\sim}{t}$.

In the local form (7.19), the equations of balance of momentum represent the set of three differential equations of motion for points of a body B.

The principle of balance of moment of momentum states that the rate of change of the moment of momentum of a part v of a body is equal to the total resultant moment of forces acting on v.

From the discussion in the section 5 we see that

the expression (5.1.25) may be considered as a general form of the moment of momentum, since various physical models which lead to continuum models yield for the moment of momentum expressions of that form. Using (6.29) we may write directly the principle of balance of moment of momentum,

$$\frac{d}{dt}\int_v \varrho(\underset{\sim}{r}\times\underset{\sim}{v} + i^{\lambda\mu}\underset{\sim}{d}_{(\lambda)}\times\dot{\underset{\sim}{d}}_{(\mu)})dv = \int_v \varrho(\underset{\sim}{r}\times\underset{\sim}{f} + \underset{\sim}{d}_{(\lambda)}\times\underset{\sim}{k}^{(\lambda)} + \underset{\sim}{l})dv + \oint_s(\underset{\sim}{r}\times\underset{\sim}{t} + \underset{\sim}{d}_{(\lambda)}\times\underset{\sim}{h}^{(\lambda)} + \underset{\sim}{m})ds \,. \tag{7.21}$$

For Cartesian coordinates by the application of (6.30) in the component form, the relation (7.21) reduces to

$$\begin{aligned}\frac{d}{dt}\int_v \varrho(z^{[\alpha}\dot{z}^{\beta]} + i^{\lambda\mu}d^{[\alpha}_{(\lambda)}\dot{d}^{\beta]}_{(\mu)})dv &= \int_v \varrho(z^{[\alpha}f^{\beta]} + d^{[\alpha}_{(\lambda)}k^{(\lambda)\beta]} + l^{\alpha\beta})dv + \\ &+ \oint_s (z^{[\alpha}t^{\beta]\gamma} + d^{[\alpha}_{(\lambda)}h^{(\lambda)\beta]\gamma} + m^{\alpha\beta\gamma})n_\gamma ds \,.\end{aligned} \tag{7.22}$$

Differentiating the integral on the left-hand side of (7.22), applying the divergence theorem to the surface integral, using the continuity equation and the equations of motion (7.18), and since the coefficients $i^{\lambda\mu}$ are symmetric, from (7.22) we obtain

$$\int_v \varrho(\dot{\overline{i^{\lambda\mu}}}\, d^{[\alpha}_{(\lambda)}\dot{d}^{\beta]}_{(\mu)} + i^{\lambda\mu}d^{[\alpha}_{(\lambda)}\ddot{d}^{\beta]}_{(\mu)})dv = \int_v [\varrho(d^{[\alpha}_{(\lambda)}k^{(\lambda)\beta]} + l^{\alpha\beta}) + t^{[\alpha\beta]} + (d^{[\alpha}_{(\lambda)}h^{(\lambda)\beta]\gamma} + m^{\alpha\beta\gamma})_{,\gamma}]dv \,. \tag{7.23}$$

However, from the analysis in the sections 5.1 and 5.2 it follows that the coefficients $i^{\lambda\mu}$ may be assumed to be independent of time, and since the relation (7.23) has to be valid for an arbitrary part v of the body, we obtain from (7.23) the local form of the principle of balance of moment of momentum,

$$\varrho i^{\lambda\mu} d^{[\alpha \cdot\cdot}_{(\lambda)} \dot{d}^{\beta]}_{(\mu)} = t^{[\beta\alpha]} + \varrho(d^{[\alpha}_{(\lambda)} k^{(\lambda)\beta]} + \ell^{\alpha\beta}) + (d^{[\alpha}_{(\lambda)} h^{(\lambda)\beta]\gamma} + m^{\alpha\beta\gamma})_{,\gamma} \,. \quad (7.24)$$

Let us introduce the notation

$$i^{\lambda\mu} d^{[\alpha}_{(\lambda)} \dot{d}^{\beta]}_{(\mu)} = \sigma^{\alpha\beta} ,$$

$$\ell^{\alpha\beta} + d^{[\alpha}_{(\lambda)} k^{(\lambda)\beta]} = \overset{*}{\ell}{}^{\alpha\beta} ,$$

$$m^{\alpha\beta\gamma} + d^{[\alpha}_{(\lambda)} h^{(\lambda)\beta]\gamma} = \overset{*}{m}{}^{\alpha\beta\gamma} , \quad (7.25)$$

$$d^{\alpha}_{(\lambda)} h^{(\lambda)\beta\gamma} = H^{\alpha\beta\gamma} ,$$

$$d^{\alpha}_{(\lambda)} k^{(\lambda)\beta} = k^{\alpha\beta} .$$

With this notation the relation (7.24) obtains the simple form

$$\varrho \dot{\sigma}^{\alpha\beta} = t^{[\beta\alpha]} + \varrho \overset{*}{\ell}{}^{\alpha\beta} + \overset{*}{m}{}^{\alpha\beta\gamma}{}_{,\gamma} \,. \quad (7.26)$$

The principle of moment of momentum in this form (for elastic materials) was obtained by Toupin [463] from Hamilton's principle. He named $\sigma^{\alpha\beta}$ the <u>spin angular momentum per unit mass</u>, $H^{\alpha\beta\gamma}$ corresponds to Toupin's <u>hyperstresses</u>, and

$H^{[\alpha\beta]\gamma}$ he identified with the couple-stress tensor. The apparent discrepancy in the terminology and symbols is due to the fact that Toupin considered separately materials with directors, and materials which are described in terms of a strain-gradient theory. The couple stress tensor $\underset{\sim}{m}$ which we introduced independently of the hyperstresses corresponds to the couple-stress tensor in Toupin's strain-gradient theory.

From (7.24) and (7.25) it is evident that it is impossible in the total effect to separate the influence of body moments from the director moments, and the influence of couple-stresses from the hyperstresses.

Assuming that there are no deformations of the directors and that there are no director forces and director stresses, the relation (7.26) reduces to

$$t^{[\alpha\beta]} = m^{\alpha\beta\gamma}{}_{,\gamma} + \varrho \ell^{\alpha\beta} , \tag{7.27}$$

which substitutes Cauchy's second law

$$t^{\alpha\beta} = t^{\beta\alpha} \tag{7.28}$$

valid only in the non-polar case.

In the theory of anisotropic fluids and liquid crystals , Ericksen [101 -117] writes a separate equation of balance for the director momentum. Ericksen considers liquid crystals as packets of rod-like molecules, which correspond to a one-director continuum model. Generalizing this idea we may introduce the principle of balance of the director moments (Stojanović, Djurić, Vujosević [428] , Djurić [86] , Stojanovič and Djurić [426] in the form

$$\frac{d}{dt}\int_v \varrho i^{\lambda\mu} \dot{d}^{\alpha}_{(\mu)} dv = \oint_S h^{(\lambda)\alpha\beta} ds_\beta + \int_v \varrho k^{(\lambda)\alpha} dv, \qquad (7.29)$$

where on the right-hand side we have written in the component form the expression for the total resultant director force (6.21)

Performing the indicated differentiation and applying the divergence theorem in (7.29) we obtain

$$\varrho i^{\lambda\mu} \ddot{d}^{\alpha}_{(\mu)} = h^{(\lambda)\alpha\beta}{}_{,\beta} + \varrho k^{(\lambda)\alpha} \qquad (7.30)$$

as an independent set of the differential equations of motion for the directors.

Using (7.30), the equations (7.24) may be reduced to the form which does not include explicitly the inertial terms,

$$t^{[\alpha\beta]} = m^{\alpha\beta\gamma}{}_{,\gamma} + \varrho l^{\alpha\beta} + d^{[\alpha}_{(\lambda),\gamma} h^{(\lambda)\beta]\gamma}, \qquad (7.31)$$

and which admits the non-vanishing of $t^{[\alpha\beta]}$ also in non-oriented media.

It is obvious that the antisymmetric part of the stress tensor is affected by the director stresses if the medium considered is an oriented medium.

Since all the equations of motion (7.18), (7.26), (7.30) are tensorial equations, we shall write these equations directly in the component form valid for an arbitrary system of curvilinear coordinates

$$\varrho \ddot{x}^i = t^{ij}{}_{,j} + \varrho f^i , \tag{7.32}$$

$$\varrho i^{\lambda\mu} \ddot{d}^i_{(\mu)} = h^{(\lambda)ij}{}_{,j} + \varrho k^{(\lambda)i} , \tag{7.33}$$

$$\varrho \dot{\sigma}^{ij} = t^{[ji]} + \varrho \overset{*}{\ell}{}^{ij} + \overset{*}{m}{}^{ijk}{}_{,k} , \tag{7.34}$$

$$(i,j,k = 1,2,3; \quad \lambda,\mu = 1,2,\ldots,n) .$$

Eliminating from (7.34) the spin angular momentum $\underset{\sim}{\sigma}$, as it was already done in (7.31), decomposing in (7.32) the stress tensor into its symmetric and antisymmetric parts and substituting the antisymmetric part from (7.34), we obtain the set of $3n + 3$ differential equations of motion,

$$\varrho \ddot{x} = t^{(ij)}{}_{,j} + m^{ijk}{}_{,jk} + (d^{[i}_{(\lambda),k} h^{(\lambda)j]k})_{,j} + \varrho \ell^{ij}{}_{,j} + \varrho f^i , \tag{7.35}$$

$$\varrho i^{\lambda\mu} \ddot{d}^i_{(\mu)} = h^{(\lambda)ij}{}_{,j} + \varrho k^{(\lambda)i} . \tag{7.36}$$

Obviously, the motion $x^i = x^i(\underset{\sim}{X},t)$ is affected by the deformations of the directors and by the director stresses, and the motion of the directors, $d^i_{(\lambda)} = d^i_{(\lambda)}(\underset{\sim}{X},t)$ is affected only by the director stresses and director forces.

It is in some cases more convenient to use the equations of motion written in the compact vectorial notation, than in the component form. If we multiply the relations (7.32) and (7.33) with the base vectors $\underset{\sim}{g}_i$ (see Appendix, Sections A1 and A3), we obtain

$$\ddot{x}^i \underset{\sim}{g}_i = \frac{d\underset{\sim}{v}}{dt} = \dot{\underset{\sim}{v}} , \quad \ddot{d}^i_{(\mu)} \underset{\sim}{g}_i = \frac{d\dot{\underset{\sim}{d}}_{(\mu)}}{dt} = \ddot{\underset{\sim}{d}}_{(\mu)} ,$$

$$t^{ij}{}_{,j} \underset{\sim}{g}_i = \frac{1}{\sqrt{g}} \partial_j (\sqrt{g}\, \underset{\sim}{t}^j) , \quad h^{(\lambda)ij}{}_{,j} \underset{\sim}{g}_i = \frac{1}{\sqrt{g}} \partial_j (\sqrt{g}\, \underset{\sim}{h}^{(\lambda)j}) ,$$

$$f^i \underset{\sim}{g}_i = \underset{\sim}{f} , \quad k^{(\lambda)i} \underset{\sim}{g}_i = \underset{\sim}{k}^{(\lambda)} ,$$

and (7.32) may be written in the form

$$\varrho \dot{\underset{\sim}{v}} = \frac{1}{\sqrt{g}} \partial_j (\sqrt{g}\, \underset{\sim}{t}^j) + \varrho \underset{\sim}{f} , \tag{7.37}$$

$$\varrho i^{\lambda\mu} \ddot{\underset{\sim}{d}}_{(\mu)} = \frac{1}{\sqrt{g}} \partial_j (\sqrt{g}\, \underset{\sim}{h}^{(\lambda)j}) + \varrho \underset{\sim}{k}^{(\lambda)} . \tag{7.38}$$

Composition of (7.34) with the Ricci alternating tensor ε_{mij} gives the vectorial equation

$$\varrho \dot{\sigma}_m = (\underset{\sim}{g}_i \times \underset{\sim}{t}^i) \cdot \underset{\sim}{g}_m + \varrho \overset{*}{\ell}_m + \frac{1}{\sqrt{g}} \partial_k (\sqrt{g}\, \overset{*}{\underset{\sim}{m}}{}^k) \cdot \underset{\sim}{g}_m , \tag{7.39}$$

where we have used (6.12) and (6.15), and

$$\sigma_m = \frac{1}{2} \varepsilon_{mij} \sigma^{ij} , \quad \overset{*}{\ell}_m = \frac{1}{2} \varepsilon_{mij} \overset{*}{\ell}{}^{ij} ,$$

$$\underset{\sim}{t}^j = t^{ij} \underset{\sim}{g}_i , \quad \overset{*}{m}{}_m^{\cdot k} = \frac{1}{2} \varepsilon_{mij} \overset{*}{m}{}^{ijk} , \tag{7.40}$$

$$\overset{*}{\underset{\sim}{m}}{}^k = \overset{*}{m}{}_m^{\cdot k} \underset{\sim}{g}^m .$$

Thus, we may write

$$\varrho \dot{\underset{\sim}{\sigma}} = \underset{\sim}{g}_k \times \underset{\sim}{t}^k + \varrho \overset{*}{\underset{\sim}{l}} + \frac{1}{\sqrt{g}} \partial_k (\sqrt{g}\, \overset{*}{\underset{\sim}{m}}{}^k) . \tag{7.41}$$

7.1 The Cosserat Continuum

The Cosserat continuum is the medium in which the directors represent rigid triads of unit vectors, so that the motion is described by the motion of points and by an independent rotation of the director triads. According to (5.4.11) the rotation of the directors is determined by the field of the angular velocity tensor $\underset{\sim}{\omega}(x,t)$, so that we have

$$\dot{d}^{i}_{(\mu)} = \omega_n^{\cdot i} d^{n}_{(\mu)} , \tag{7.1.1}$$

from which follows

$$\ddot{d}^{i}_{(\mu)} = (\dot{\omega}_l^{\cdot i} + \omega_l^{\cdot n} \omega_n^{\cdot i}) d^{l}_{(\mu)} . \tag{7.1.2}$$

The angular velocity tensor $\underset{\sim}{\omega}$ is antisymmetric and instead of nine functions $d^{i}_{(\mu)}(\underset{\sim}{x},t)$ we have to consider only three independent components of $\underset{\sim}{\omega}$.

From (7.34) and (7.25) we easily obtain three independent equations for the determination of the angular velocity tensor,

$$\varrho [I^{li}(\dot{\omega}_l^{\cdot j} + \omega_l^{\cdot n} \omega_n^{\cdot j})]_{[ij]} = t^{[ij]} + \overset{*}{m}{}^{ijk}{}_{,k} + g \overset{*}{l}{}^{ij} , \tag{7.1.3}$$

where

$$I^{ij} = I^{ji} = i^{\lambda\mu} d^i_{(\lambda)} d^j_{(\mu)} , \qquad (7.1.4)$$

which represent the density of inertia coefficients per unit mass

According to (5.1.10), for a particle consisting of n mass points the directors $\underset{\sim}{d}_{(\lambda)}$ are position vectors of the mass points with respect to the centres of mass of corresponding particles, and therefore we have

$$I^{\alpha\beta} = \frac{1}{m}\sum_{\nu=1}^{n} m_\nu \delta^\pi_\nu \delta^\varrho_\nu \xi^\alpha_{(\pi)} \xi^\beta_{(\varrho)} = \frac{1}{m}\sum_{\nu=1}^{n} m_\nu \xi^\alpha_{(\nu)} \xi^\beta_{(\nu)} .$$

Hence, $I^{\alpha\beta}$ are components of the inertia tensor of the particle considered. Also for the media with microstructure when a curvilinear system of coordinates x^i is introduced into (5.2.32) and (5.2.36) and when the vectors $\underset{\sim}{\chi}_\alpha$ are identified with the directors, a relation of the form of (7.1.4) will be obtained.

Taking the material derivative of I^{ij} (with $i^{\lambda\mu}$ independent of time) and using (7.1.1) we find

$$\frac{\partial I^{ij}}{\partial t} + I^{ij}_{\ ,k} v^k - I^{kj} \omega_k^{\ i} - I^{ik} \omega_k^{\ j} = 0 . \qquad (7.1.5)$$

This relation Eringen [124] calls the conservation of micro--inertia.

The complete set of equations of motion of a Cosserat continuum consists now of the following equations

1) $$\frac{\partial \varrho}{\partial t} + (\varrho v^k)_{,k} = 0 , \qquad (7.1.6)$$

$$\rho\ddot{x}^i = t^{ij}{}_{,j} + \rho f^i , \qquad (2$$

(7.1.6) $$\dot{I}^{ij} - I^{kj}\omega_k^{\cdot i} - I^{ik}\omega_k^{\cdot j} = 0 , \qquad (3$$

$$\rho[I^{li}(\dot{\omega}_l^{\cdot j} + \omega_l^{\cdot n}\omega_n^{\cdot j})]_{[ij]} = t^{[ji]} + \overset{*}{m}{}^{ijk}{}_{,k} + \rho\overset{*}{l}{}^{ij} . \qquad (4$$

Substituting in the last equation (7.1.6) the angular velocity tensor ω^{ij} by the angular velocity vector $\boldsymbol{\omega}$,

$$\omega_m = \frac{1}{2}\varepsilon_{mij}\omega^{ij}$$

and recalling (7.41), we may write (7.1.6) in the form

(7.1.7) $$\rho\dot{\boldsymbol{\sigma}} = \rho\boldsymbol{j}\cdot\dot{\boldsymbol{\omega}} + \rho(\boldsymbol{I}\cdot\boldsymbol{\omega})\times\boldsymbol{\omega} = \boldsymbol{g}_k\times\boldsymbol{t}^k + \rho\overset{*}{\boldsymbol{l}} + \frac{1}{\sqrt{g}}\partial_k(\sqrt{g}\,\overset{*}{\boldsymbol{m}}{}^k),$$

where we have put

(7.1.8) $$\boldsymbol{j} = \{j^m_{\cdot n}\} = \{I^t_t\delta^m_n - I^m_n\} ,$$

(7.1.9) $$\boldsymbol{I}\cdot\boldsymbol{\omega} = \{I^m_n\omega^n\} .$$

In the linear theories it is assumed that the angular velocity is sufficiently small, such that the spin moment may be approximated by

$$\dot{\boldsymbol{\sigma}} = \boldsymbol{j}\cdot\dot{\boldsymbol{\omega}} .$$

For microisotropic materials (cf. Eringen [132]) it is assumed that

$$\underset{\sim}{j} = j\underset{\sim}{1}\,, \quad \text{i.e.} \quad j^m_n = j\delta^m_n\,. \qquad (7.1.10)$$

A very interesting field of application of the theory of Cosserat media is the dynamics of gradual media. Oshima [347] considered a model of a granular medium assuming that there are no director forces and director stresses and disregarding the coefficients of inertia of the granulae. Cowin [74] assumes the same kinematical model as Oshima. A more general approach is offered by the theory of micropolar media (Eringen [123 - 127]), but this theory is not yet explicitly applied to granular materials. Satake considered first [385] a granular medium in the absence of volume and director forces and moments, but in a recent paper [386] he included these forces into the consideration. Satake approaches the problem from the point of view of a purely linear theory and, the same as Oshima, he assumes certain a priori described mechanical properties of the medium (elasticity). Cowin admits the medium to be a composition of elastic and viscous phases.

A much wider field of applications is offered if the directors do not constitute rigid trihedra. The micropolar theory of Eringen generalizes the idea of a Cosserat continuum admitting the directors to deform, but restricting the number of directors to three. A large number of applications is covered by the later development of the micropolar theory. (Cf. e.g. Ariman [14,15] , Ariman and Cakmak [18], Ariman, Cakmak and Hill [17] , Askar and Cakmak [19] , Askar, Cakmak and Ariman [20]).

A structural model of a micropolar continuum (Askar and Cakmak [19]), which consists of a two-dimensional network of orientable points, joined by extensible and flexible points, yields the equations very close to those obtained by Eringen and Suhubi [138, 442] , Eringen [126] and Mindlin [286, 291] , starting with continuum principles.

7.2 Bodies with One Director

The theory of liquid crystals and anisotropic fluids of Ericksen [101-117] (cf. also Leslie [267, 268]) is based on the assumption that the media such as liquid crystals and suspensions of large molecules may be described by the position vectors of the particles and by a simple director field. The differential equations of motion may be obtained from our equations (7.32-34), together with the continuity equation (7.11):

$$\dot{\varrho} + \varrho v^k_{,k} = 0 ,$$

$$\varrho \ddot{x}^i = t^{ij}_{,j} + \varrho f^i , \tag{7.2.1a}$$

$$\ddot{d}^i = k^i ,$$

$$t^{[ij]} = k^{[i} d^{j]} . \qquad (7.2.1b)$$

To obtain these equations from (7.11) and (7.25) we have to assume that there are no director stresses $\underset{\sim}{h}$, no couple-stresses $\underset{\sim}{m}$ and no volume couples $\underset{\sim}{l}$. Under such assumptions the equation $(7.2.1)_4$ is a direct consequence of the moment of momentum equation (7.24).

Another example of a one-director theory is the theory of <u>Cosserat surfaces</u>. (Green, Naghdi and Wainwright [171] Green and Naghdi [163 - 167]).

A Cosserat surface is a two-dimensional material manifold s to each point of which a simple director field is assigned. This surface is embedded in a three-dimensional Euclidean space. Let $x^\alpha, \alpha = 1,2$ be coordinates defining points on the surface and $x^3=0$ at all points of the surface. The position vector of a point of s at time t and the director $\underset{\sim}{d}$ are functions of position x^α and of time t,

$$\underset{\sim}{r} = \underset{\sim}{r}(x^\alpha, t) , \quad \underset{\sim}{d} = \underset{\sim}{d}(x^\alpha, t) . \qquad (7.2.2)$$

The base vectors along curves x^α are $\underset{\sim}{g}_\alpha$ and we assume that $\underset{\sim}{g}_3$ is the unit normal vector to s, so that

$$\begin{aligned} &\underset{\sim}{g}_\alpha \cdot \underset{\sim}{g}_\beta = g_{\alpha\beta} , \quad (\underset{\sim}{g}_\alpha \times \underset{\sim}{g}_\beta) \cdot \underset{\sim}{g}_3 > 0 , \quad (\alpha \neq \beta) \\ &\underset{\sim}{g}^\alpha \cdot \underset{\sim}{g}_\beta = \delta^\alpha_\beta , \quad \underset{\sim}{g}^3 \cdot \underset{\sim}{g}_\alpha = 0 , \quad \underset{\sim}{g}_3 \cdot \underset{\sim}{g}_3 = 1 \end{aligned} \qquad (7.2.3a)$$

(7.2.3b) $$\underset{\sim}{g}^{\alpha} = g^{\alpha\beta}\underset{\sim}{g}_{\beta} \, .$$

From the theory of surfaces it is known that the second fundamental tensor $b_{\alpha\beta}$ of a surface is defined by

(7.2.4) $$\underset{\sim}{g}_{\alpha|\beta} = b_{\alpha\beta}\underset{\sim}{g}_3 \, , \qquad \frac{\partial \underset{\sim}{g}_3}{\partial x^{\beta}} = -b^{\alpha}_{\beta}\underset{\sim}{g}_{\alpha} \, ,$$

where "|" denotes covariant differentiation with respect to the metric form on the surface s.

Let $\underset{\sim}{F}$ and $\underset{\sim}{k}$ be the assigned force and the assigned director force per unit mass,

(7.2.5) $$\underset{\sim}{F} = F^{\alpha}\underset{\sim}{g}_{\alpha} + F^{3}\underset{\sim}{g}_3$$

$$\underset{\sim}{k} = k^{\alpha}\underset{\sim}{g}_{\alpha} + k^{3}\underset{\sim}{g}_3 \, .$$

The stress vector $\underset{\sim}{t}^{\alpha}$ is to be regarded as a force per unit length of a curve bounding an area on s . The same holds for the director stress $\underset{\sim}{h}^{\alpha}$, so that

(7.2.6) $$\underset{\sim}{t}^{\alpha} = t^{\beta\alpha}\underset{\sim}{g}_{\beta} + t^{3\alpha}\underset{\sim}{g}_3 \, ,$$

$$\underset{\sim}{h}^{\alpha} = h^{\beta\alpha}\underset{\sim}{g}_{\beta} + h^{3\alpha}\underset{\sim}{g}_3 \, .$$

To write the equation of continuity (7.11) in the appropriate form we have to calculate the divergence of the velocity vector $\underset{\sim}{v}$ considering (7.2.3). Let the velocity vector of a point on s be

$$\underset{\sim}{v} = v^{\alpha}\underset{\sim}{g}_{\alpha} + v^{3}\underset{\sim}{g}_3 \, .$$

The Hamiltonian operator on the surface s is

$$\underset{\sim}{\nabla} = \underset{\sim}{g}^{\alpha}\partial_{\alpha}$$

and we have

$$v^i_{,i} = \underset{\sim}{\nabla}\cdot\underset{\sim}{v} = \underset{\sim}{g}^{\alpha}\cdot(\underset{\sim}{g}_{\beta}\partial_{\alpha}v^{\beta} + v^{\beta}\partial_{\alpha}\underset{\sim}{g}_{\beta} + \underset{\sim}{g}_3\partial_{\alpha}v^3 + v^3\partial_{\alpha}\underset{\sim}{g}_3) ,$$

which in virtue of (7.2.4) becomes

$$v^i_{,i} = v^{\alpha}{}_{|\alpha} - b^{\alpha}_{\alpha}v^3 .$$

Substituting this in (7.11) we obtain the continuity equation in the form

$$\dot{\varrho} + \varrho(v^{\alpha}{}_{|\alpha} - b^{\alpha}_{\alpha}v^3) = 0 . \qquad (7.2.7)$$

Differentiation of the stress vectors $\underset{\sim}{t}^{\alpha}$ gives

$$\underset{\sim}{t}^{\alpha}{}_{,\gamma} = t^{\beta\alpha}{}_{|\gamma}\underset{\sim}{g}_{\beta} + t^{\beta\alpha}\underset{\sim}{g}_{\beta|\gamma} + t^{3\alpha}{}_{|\gamma}\underset{\sim}{g}_3 + t^{3\alpha}\underset{\sim}{g}_{3|\gamma} ,$$

which because of (7.2.4), reduces to

$$\underset{\sim}{t}^{\alpha}{}_{,\gamma} = (t^{\beta\alpha}{}_{|\gamma} - b^{\beta}_{\gamma}t^{3\alpha})\underset{\sim}{g}_{\beta} + (t^{3\alpha}{}_{|\gamma} + b_{\beta\gamma}t^{\beta\alpha})\underset{\sim}{g}_3 . \qquad (7.2.8)$$

We obtain the similar expression for the derivatives of the director stress vectors $\underset{\sim}{h}^{\alpha}$,

$$h^{\alpha}{}_{,\gamma} = (\underset{\sim}{h}^{\beta\alpha}{}_{|\gamma} - b^{\beta}_{\gamma}h^{3\alpha})\underset{\sim}{g}_{\beta} + (h^{3\alpha}{}_{|\gamma} + b_{\beta\gamma}h^{\beta\alpha})\underset{\sim}{g}_3 . \qquad (7.2.9)$$

From the vectorial form of the differential equations of motion (7.19),

$$\varrho \dot{\underset{\sim}{v}} = \underset{\sim}{t}^{i}_{;i} + \varrho \underset{\sim}{f} ,$$

by scalar multiplication with the base vectors $\underset{\sim}{g}^{\alpha}$ and $\underset{\sim}{g}^{3}$ we obtain the following three differential equations of motion:

$$(7.2.10) \qquad \varrho a^{\beta} = t^{\beta\alpha}{}_{|\alpha} - b^{\beta}_{\alpha} t^{3\alpha} + \varrho f^{\beta} ,$$

$$\varrho a^{3} = t^{3\alpha}{}_{|\alpha} + b_{\alpha\beta} t^{\beta\alpha} + \varrho f^{3} ,$$

where $\underset{\sim}{a}$ is the acceleration vector with the components (a^{α}, a^{3}) .

Green, Naghdi and Wainwright [171] assumed that there is an additional physical director force which they denoted by $\underset{\sim}{m}^{\alpha}$ and which acts over the curves x^{α}.

For the motion of the director $\underset{\sim}{d}(\underset{\sim}{x}^{\alpha},t)$ we shall write also the equations (7.33) in the compact (vectorial) form to which our equations (7.33) reduce in the case of a single director field,

$$\underset{\sim}{m} = \underset{\sim}{h}^{i}_{;i} + \varrho(\underset{\sim}{k} - i\ddot{\underset{\sim}{d}}) ,$$

where $\underset{\sim}{m}$ represents the additional physical force, and ϱi is the inertia density at the points of the surface. Since the director stress depends only upon x^{α}, we may write

$$(7.2.11) \qquad \underset{\sim}{m} = \underset{\sim}{h}^{\alpha}{}_{|\alpha} + \varrho(k - i\ddot{\underset{\sim}{d}}) ,$$

and by scalar multiplication with $\underset{\sim}{g}^{\beta}$ and $\underset{\sim}{g}^{3}$ this equation gives the following equations in the component form:

$$m^{\beta} = h^{\beta\alpha}{}_{|\alpha} - b^{\beta}_{\alpha} h^{3\alpha} + \varrho(k^{\beta} - i\ddot{d}^{\beta}) ,$$
$$m^{3} = h^{3\alpha}{}_{|\alpha} + b_{\alpha\beta} h^{\beta\alpha} + \varrho(k^{3} - i\ddot{d}^{3}) . \qquad (7.2.12)$$

The equations (7.2.7), (7.2.10) and (7.2.12) represent the basic set of equations for a Cosserat surface. In the original paper of Green, Naghdi and Wainwright, as well as in the subsequent work of Green and Naghdi, the equations of motion are derived directly from the considerations of the surface, and not from a general theory of the generalized Cosserat continua.

In the applications of the theory of Cosserat surfaces to the theory of elastic plates and shells it was assumed that in the initial configuration $\underset{\sim}{D}_{(\alpha)} = 0$ and $\underset{\sim}{D}_{(3)} = e_3$. For further references see e.g. [163, 164, 167, 318] . *

---

* Ericksen and Truesdell [121] gave a very elegant and exact theory of strain and stress in shells, assuming that three directors are assigned to each point of the surface. The work of Cohen and DeSilva [64, 65] on elastic surfaces is based also on the assumption that three directors are assigned to the points of the surface, and they based their work on the results of Ericksen and Truesdell. Their equations of equilibrium may be derived directly from our equations (7.32, 33). However, in the theory of elastic membranes [66] they consider, at the points of the membrane, a single director field. The director is taken to be normal to the surface and the only deformation it suffers is the deformation of its magnitude.

7.3 Bodies with Two Directors
A Theory of Rods

As an example of two-director bodies we shall consider the theory of rods by Green and Laws [153, 155], which was applied to the theory of elastic rods by Green, Naghdi and Laws [156].

A rod is considered as a curve ℓ, imbedded in Euclidean three-dimensional space. At each point of the curve there are two assigned directors. Let θ be a convected coordinate * defining points on the curve, and let $\underset{\sim}{r}$ be the position vector, relative to a fixed origin, of a point on the curve,

$$\underset{\sim}{r} = \underset{\sim}{r}(\theta,t) . \tag{7.3.1}$$

Let $\underset{\sim}{d}_{(1)} = \underset{\sim}{g}_1$ and $\underset{\sim}{d}_{(2)} = \underset{\sim}{g}_2$ be the assigned directors and let the vector $\underset{\sim}{g}_3$ tangential to the curve,

$$\underset{\sim}{g}_3 = \frac{\partial \underset{\sim}{r}}{\partial \theta} , \tag{7.3.2}$$

be considered as the third vector of the triad, so that

$$g = (\underset{\sim}{g}_1 \times \underset{\sim}{g}_2)\cdot\underset{\sim}{g}_3 > 0 .$$

---

* Convected coordinates, by the definition, move with the body and deform with it so that the numerical values of such coordinates for each point of the body remain unchanged.

Along ℓ we may construct the reciprocal triad $\underset{\sim}{g}^i$, such that

$$\underset{\sim}{g}_i \underset{\sim}{g}_j = g_{ij}, \quad \underset{\sim}{g}^i \underset{\sim}{g}^j = g^{ij}, \quad g^{ij}\underset{\sim}{g}_j = \underset{\sim}{g}^i \tag{7.3.3}$$

$$\underset{\sim}{g}^i \underset{\sim}{g}_j = \delta^i_j, \quad g^{ij} g_{jk} = \delta^i_k .$$

We shall introduce the notation

$$\frac{\partial \underset{\sim}{g}_i}{\partial \theta} \cdot \underset{\sim}{g}_j = x_{ij}, \quad g^{jk} \cdot x_{ij} \equiv \underset{\sim}{g}^k \cdot \frac{\partial \underset{\sim}{g}_i}{\partial \theta} = x_i^{\cdot k} . \tag{7.3.4}$$

It is assumed that the stress acts along the curve ℓ. The stress vector $\underset{\sim}{t}(\theta,\underset{\sim}{n})$ according to (6.11) is

$$\underset{\sim}{t}(\theta,\underset{\sim}{n}) = t^i(\theta,\underset{\sim}{n})\underset{\sim}{g}_i = t^{i3}\underset{\sim}{g}_i n_3 \equiv t^i \underset{\sim}{g}_i . \tag{7.3.5}$$

Since $n_3 = n = 1$, the components of the stress tensor reduce to $t^{i3} = t^i$ the total resultant stress exerted on a segment (θ_1,θ_2) of a rod is

$$\underset{\sim}{t}(\theta_2) - \underset{\sim}{t}(\theta_1) = \left[\underset{\sim}{t}(\theta)\right]_{\theta_1}^{\theta_2} . \tag{7.3.6}$$

For the director stress vectors $h^{(\alpha)}$, according to (6.19) and (6.20) we may also write

$$\underset{\sim}{h}^{(\alpha)}(\theta,\underset{\sim}{n}) = h^{(\alpha)i}(\theta,\underset{\sim}{n})\underset{\sim}{g}_i = h^{(\alpha)i3}(\theta)\underset{\sim}{g}_i \cdot n_3 = h^{(\alpha)i}\underset{\sim}{g}_i , \tag{7.3.7}$$

and the moment of the director stresses, defined by (6.27), becomes

$$\underset{\sim}{\mu} \equiv \underset{\sim}{d}_{(\alpha)} \times h^{(\alpha)i}\underset{\sim}{g}_i = h^{(\alpha)i}\underset{\sim}{g}_\alpha \times \underset{\sim}{g}_i . \tag{7.3.8}$$

The resultant moment of the director stresses exerted on the segment (Θ_1, Θ_2) of the rod will be according to (6.28),

$$\underset{\sim}{\mu}(\Theta_2) - \underset{\sim}{\mu}(\Theta_1) = \left[\underset{\sim}{\mu}(\Theta)\right]_{\Theta_1}^{\Theta_2} . \tag{7.3.9}$$

If we assume that there are no body couples $\underset{\sim}{l}$ and no couple stresses $\underset{\sim}{m}$ acting on the curve l, and since the mass dm of the line element ds is given by

$$dm = \varrho ds = \varrho \sqrt{g_{33}}\, d\Theta , \tag{7.3.10}$$

the law of conservation of mass and the principles of balance of momentum (7.14) and of the moment of momentum (7.21) obtain the form

$$\frac{d}{dt}\int_{\Theta_1}^{\Theta_2} \varrho \sqrt{g_{33}}\, d\Theta = 0 , \tag{7.3.11}$$

$$\frac{d}{dt}\int_{\Theta_1}^{\Theta_2} \varrho \underset{\sim}{v} \sqrt{g_{33}}\, d\Theta = \int_{\Theta_1}^{\Theta_2} \varrho \underset{\sim}{f} \sqrt{g_{33}}\, d\Theta + \left[\underset{\sim}{t}(\Theta)\right]_{\Theta_1}^{\Theta_2} , \tag{7.3.12}$$

$$\frac{d}{dt}\int_{\Theta_1}^{\Theta_2} \varrho (\underset{\sim}{r} \times \underset{\sim}{v} + i^{\lambda\mu} \underset{\sim}{d}_{(\lambda)} \times \dot{\underset{\sim}{d}}_{(\mu)}) \sqrt{g_{33}}\, d\Theta = \tag{7.3.13}$$

$$= \int_{\Theta_1}^{\Theta_2} \varrho (\underset{\sim}{r} \times \underset{\sim}{f} + \underset{\sim}{d}_{(\lambda)} \times \underset{\sim}{k}^{(\lambda)}) \sqrt{g_{33}}\, d\Theta + \left[\underset{\sim}{r} \times \underset{\sim}{t} + \underset{\sim}{\mu}\right]_{\Theta_1}^{\Theta_2} .$$

Since Θ_1 and Θ_2 are convected coordinates of two points of the curve and remain unchanged under the deformations of the curve, it follows from (7.3.11) that $\varrho\sqrt{g_{33}}$ is independent of time and the law of conservation of mass may be written in the form

$$\varrho\sqrt{g_{33}} = J(\Theta) , \tag{7.3.14}$$

where $J(\Theta)$ is an arbitrary function of position.

Using the simple relation

$$\left[f(\Theta)\right]_{\Theta_1}^{\Theta_2} = \int_{\Theta_1}^{\Theta_2} \frac{df(\Theta)}{d(\Theta)}\, d\Theta ,$$

the equations (7.3.12) and (7.3.13) obtain the form*

$$\int_{\Theta_1}^{\Theta_2} \varrho \dot{\underset{\sim}{v}} \sqrt{g_{33}}\, d\Theta = \int_{\Theta_1}^{\Theta_2} \left(\varrho \underset{\sim}{f} \sqrt{g_{33}} + \frac{\partial \underset{\sim}{t}}{\partial \Theta}\right) d\Theta , \tag{7.3.15}$$

$$\int_{\Theta_1}^{\Theta_2} \varrho(\underset{\sim}{r} \times \dot{\underset{\sim}{v}} + i^{\lambda\mu} \underset{\sim}{d}_{(\lambda)} \times \ddot{\underset{\sim}{d}}_{(\mu)}) \sqrt{g_{33}}\, d\Theta = \tag{7.3.16}$$

$$= \int_{\Theta_1}^{\Theta_2} \left[\varrho(\underset{\sim}{r} \times \underset{\sim}{f} + \underset{\sim}{d}_{(\lambda)} \times \underset{\sim}{k}^{(\lambda)}) \sqrt{g_{33}} + \frac{\partial}{\partial \Theta}(\underset{\sim}{r} \times \underset{\sim}{t} + \underset{\sim}{\mu})\right] d\Theta .$$

These two equations must be valid for an arbitrary segment (Θ_1, Θ_2), which yields the local form of the equations of balance, i.e. we get the equations of motion:

$$\varrho \dot{\underset{\sim}{v}} = \varrho \underset{\sim}{f} + \frac{1}{\sqrt{g_{33}}} \frac{\partial \underset{\sim}{t}}{\partial \Theta} , \tag{7.3.17}$$

---

* We take $i^{\lambda\mu}$ to be indipendent of time [155].

$$\varrho i^{\lambda\mu} \underset{\sim}{d}_{(\lambda)} \times \ddot{\underset{\sim}{d}}_{(\mu)} = \varrho \underset{\sim}{\Gamma} + \frac{1}{\sqrt{g_{33}}} \left(\frac{\partial \underset{\sim}{\mu}}{\partial \theta} + \underset{\sim}{g}_3 \times \underset{\sim}{t} \right), \tag{7.3.18}$$

$$(\underset{\sim}{\Gamma} \equiv \underset{\sim}{d}_{(\lambda)} \times \underset{\sim}{k}^{(\lambda)}).$$

where we have applied (7.3.17) to simplify the equation (7.3.18).

To write the equations of motion in the component form we have to apply the formula

$$\frac{\partial \underset{\sim}{T}}{\partial \theta} = \left(\frac{\partial T^i}{\partial \theta} + T^m x_m^{\cdot i} \right) \underset{\sim}{g}_i ,$$

where $\underset{\sim}{T}(\theta, t)$ is a tensor defined along the curve ℓ, and $x_m^{\cdot i}$ is defined by (7.3.4). Hence, the scalar products of the vectorial equations (7.3.17, 18) with the base vectors $\underset{\sim}{g}^i$ give the following six differential equations of motion:

$$\varrho \dot{\underset{\sim}{v}} \cdot \underset{\sim}{g}^i = \varrho f^i + \frac{1}{\sqrt{g_{33}}} \left(\frac{\partial t^i}{\partial \theta} + x_m^{\cdot i} t^m \right), \tag{7.3.19}$$

$$\varrho i^{\lambda\mu} (\underset{\sim}{d}_{(\lambda)} \times \ddot{\underset{\sim}{d}}_{(\mu)}) \cdot \underset{\sim}{g}^i = \varrho \Gamma^i + \frac{1}{\sqrt{g_{33}}} \left[\frac{\partial \mu^i}{\partial \theta} + x_m^{\cdot i} \mu^m + (\underset{\sim}{g}^i \times \underset{\sim}{g}_3) \underset{\sim}{t} \right]. \tag{7.3.20}$$

Since we have

$$(\underset{\sim}{g}^i \times \underset{\sim}{g}_3) \cdot \underset{\sim}{t} = g^{ij} (\underset{\sim}{g}_j \times \underset{\sim}{g}_3) \cdot \underset{\sim}{t} = g^{ij} \varepsilon_{j3k} \underset{\sim}{g}^k \underset{\sim}{t} = g^{ij} \varepsilon_{j3k} t^k ,$$

and k must be different from 3 according to the definition of the $\underset{\sim}{\varepsilon}$-tensors, the equation (7.3.20) may also be written in the form

$$\varrho i^{\lambda\mu} (\underset{\sim}{d}_{(\lambda)} \times \ddot{\underset{\sim}{d}}_{(\mu)}) = \varrho \Gamma^i + \frac{1}{\sqrt{g_{33}}} \left(\frac{\partial \mu^i}{\partial \theta} + x_m^{\cdot i} \mu^m + g^{ij} \varepsilon_{j3\alpha} t^\alpha \right). \tag{7.3.21}$$

The equations (7.3.14), (7.3.19) and (7.3.21) represent the basic set of the equations of motion in the general theory of rods by Green and Laws.

Ericksen and Truesdell [121] assigned to each point of a rod three directors and discussed in detail the state of strain and stress from this point of view, without making any constitutive assumptions on the mechanical properties of the material of the rod. In their criticism of the classical description of the strain in a rod, the inadequacy of the classical description of twist and the insufficiencies of the theories which do not assume the material to be oriented in the sense of the generalized Cosserat continuum become obvious. Cohen's theory [63] of elastic rods is based on the kinematics and statics of Ericksen and Truesdell. An independent approach to the theory of rods, but with the same form of the equations of motion (7.3.14 19, 21) is presented by Suhubi [440].

8. Some Applications of Classical Thermodynamics

During the last ten years a great work has been done on the development of thermodynamics of continua. Our interest here is primarily directed towards the application of thermodynamics in the derivation of the constitutive equations, and we shall restrict our considerations to the classical formulations of the first law and the second law of thermodynamics. The readers interested in the modern contributions up to 1965, may be referred to the book by Truesdell and Noll [468] , and for the later work to the papers by e.g. Chen [61], Green and Laws [154], Green and Rivlin [176], Kline and Allen [236], Leigh[265], Truesdell [466, 467] , Uhlhorn [470]etc.

The experience shows that mechanical processes cannot be separated from thermal phenomena. Mechanical work may make a body hotter, or heating may produce certain mechanical effects, such as e.g. thermal dilatations and thermoelastic stresses.

To indicate how hot is a body the temperature θ is introduced as a fundamental entity. It is assumed that there exists an absolute zero $\theta = 0$ which is the lowest bound of θ and for all processes $\theta > 0$.

It is postulated that the total energy of a body is the sum of the kinetic energy produced by the motion of the mass points of the body and of an internal energy E.

For the internal energy it is assumed that it is an absolutely continuous function of mass, so that for a part v of a body it may be written

$$(8.1) \qquad E = \int_v \varepsilon \, dm = \int_v \rho \varepsilon \, dv ,$$

where ε is the specific internal energy,

$$\varepsilon = \varepsilon(\underset{\sim}{x}, t) . \tag{8.2}$$

The increment of the total energy per unit time depends on the rate P at which the mechanical forces do work (the mechanical working), and on the total input (output) of the non-mechanical working (heat), which we shall denote by Q.

Mechanical working is the rate at which the body forces $\underset{\sim}{f}$, the director forces $\underset{\sim}{k}^{(\lambda)}$, the body couples $\underset{\sim}{\ell}$, the stresses $\underset{\sim}{t}$, the director stresses $\underset{\sim}{h}^{(\lambda)}$ and couple stresses $\underset{\sim}{m}$ do work. According to the definitions of the section 6, $\underset{\sim}{t}$, $\underset{\sim}{h}^{(\lambda)}$ and $\underset{\sim}{m}$ are defined for the points on the boundary s of a part v of the body considered. Therefore the working of $\underset{\sim}{f}$, $\underset{\sim}{k}^{(\lambda)}$ and $\underset{\sim}{\ell}$ is to be summed over all the points of v, and the working of the forces $\underset{\sim}{t}$, $\underset{\sim}{h}^{(\lambda)}$ and $\underset{\sim}{m}$ over the points on the bounding surface s.

The kinetic energy T of a part v of a body we shall assume to be in the general case represented by the expression of the form (5.1.26),

$$T = \frac{1}{2}\int_v \varrho(\dot{x}^i \dot{x}_i + i^{\lambda\mu} \dot{d}^{\,i}_{(\lambda)} \dot{d}_{(\mu)i}) dv , \tag{8.3}$$

where we assume that the coefficients $i^{\lambda\mu}$ are independent of time. The rate of the kinetic energy will be now

$$\dot{T} = \int_v \varrho(\ddot{x}^i \dot{x}_i + i^{\lambda\mu} \ddot{d}^{\,i}_{(\lambda)} \dot{d}_{(\mu)i}) dv . \tag{8.4}$$

Using the equations of motion (7.32, 33, 34),

$$\varrho \ddot{x}^i = t^{ij}_{\ ;j} + \varrho f^i , \tag{8.5a}$$

(8.5b) $$\varrho i^{\lambda\mu}\ddot{d}^{i}_{(\mu)} = h^{(\lambda)ij}{}_{,j} + \varrho k^{(\lambda)i} ,$$

$$\varrho\dot{\sigma}^{ij} = t^{[ji]} + \varrho \overset{*}{l}{}^{ij} + \overset{*}{m}{}^{ijk}{}_{,k} ,$$

where owing to the tensorial character of the quantities involved we may from (7.25) write the corresponding expressions for curvilinear coordinates,

(8.6) $$\sigma^{ij} = i^{\lambda\mu} d^{[i}_{(\lambda)} \dot{d}^{j]}_{(\mu)} ,$$

$$\overset{*}{l}{}^{ij} = l^{ij} + d^{[i}_{(\lambda)} k^{(\lambda)j]} , \quad \overset{*}{m}{}^{ijk} = m^{ijk} + d^{[i}_{(\lambda)} h^{(\lambda)j]k} ,$$

and for the rate of the kinetic energy we have the expression

(8.7) $$\dot{T} = \oint_S (t^{ik}\dot{x}_i + h^{(\lambda)ik}\dot{d}_{(\lambda)i} - m^{ijk} w_{ij}) ds_k +$$
$$+ \int_v \varrho (f^i \dot{x}_i + k^{(\lambda)i}\dot{d}_{(\lambda)i} - l^{ij} w_{ij}) dv - W .$$

By W we have denoted here

(8.8) $$W = \int_v w\, dv = \int_v (t^{(ij)} d_{ij} + h^{(\lambda)jk} d^{i}_{(\lambda),k} w_{ij} + h^{(\lambda)jk}\dot{d}_{(\lambda)j,k} - m^{ijk} w_{ij,k}) dv.$$

The right-hand side of (8.7) represents the mechanical working P .

The non-mechanical working Q is assumed to rise from surface and volume densities,

(8.9) $$Q = \oint_S q^k ds_k + \int_v h\, dm ,$$

where $\underset{\sim}{q}$ is the rate at which heat flows through the surface, and h is the heat generation per unit mass (source). $\underset{\sim}{q}$ is often called the heat flux vector.

The first law of thermodynamics postulates that

$$\dot{T} + \dot{E} = P + Q . \tag{8.10}$$

From (8.10), using (8.1) and (8.5-9), we obtain

$$\varrho\dot{\varepsilon} = w + q^k_{,k} + \varrho h , \tag{8.11}$$

which represents the local law of balance and energy. According to (8.1), (8.3), (8.8) and (8.9) we see that the first law of thermodynamics is also of the form of a balance law, and therefore it represents in the global form (8.10) the law of balance of the total energy.

From experience we know that at least one part of the mechanical working goes into heat, and the rest is again available for the mechanical work. Therefore we assume that W may be decomposed into a reversible part ${}_EW$ and into an irreversible part ${}_DW$ which may also be called the dissipative part of W, such that

$$W = {}_EW + {}_DW . \tag{8.12}$$

The reversible part of working goes into the potential energy Σ, such that $\dot{\Sigma} = {}_DW$ and

$$\Sigma = \int_v {}_Ew\, dv = \int_v \varrho\sigma\, dv , \tag{8.13}$$

where σ is the specific strain energy, or the elastic potential.

The difference between the rate of the specific internal energy and the rate of reversible work we shall denote by $\theta\dot{\eta}$, so that

$$\varrho\dot{\varepsilon} = {}_E w + \varrho\theta\dot{\eta} , \tag{8.14}$$

where η represents the specific entropy and is defined per unit mass and per unit temperature, and from (8.11) we obtain

$$\varrho\theta\dot{\eta} = {}_D w + q^k_{,k} + \varrho h , \tag{8.15}$$

which represents the equation of production of specific entropy.

If we assume that all stresses, director stresses and stress-couples may be decomposed into parts which do reversible work (${}_E\underset{\sim}{t}$, ${}_E\overset{(\lambda)}{\underset{\sim}{h}}$, ${}_E\underset{\sim}{m}$), and which do dissipative work (${}_D\underset{\sim}{t}$, ${}_D\overset{(\lambda)}{\underset{\sim}{h}}$, ${}_D\underset{\sim}{m}$), we may write

$$\underset{\sim}{t} = {}_E\underset{\sim}{t} + {}_D\underset{\sim}{t} ; \quad \overset{(\lambda)}{\underset{\sim}{h}} = {}_E\overset{(\lambda)}{\underset{\sim}{h}} + {}_D\overset{(\lambda)}{\underset{\sim}{h}} ; \quad \underset{\sim}{m} = {}_E\underset{\sim}{m} + {}_D\underset{\sim}{m}. \tag{8.16}$$

From (8.15) it follows that any portion of the stress, director stresses and couple-stresses which does recoverable work makes no contribution to the entropy (Truesdell and Toupin [469]).

On the basis of (8.11, 12, and 15) we may write

$$\int_v \varrho\dot{\eta}\,dv - \int_v \left(\frac{1}{\theta}q^k_{,k} + \frac{\varrho}{\theta}h\right)dv = \int_v \frac{1}{\theta}\,{}_D W\,dv . \tag{8.17}$$

The quantity H defined by

$$H = \int_v \varrho\eta dv$$

is called the total entropy. Now, from (8.17) we obtain

$$\dot{H} - \oint_S \frac{q^k ds_k}{\theta} - \int_v \frac{\varrho h}{\theta} dv = \int_v \frac{1}{\theta}\left({}_D W + \frac{\theta_{,k} q^k}{\theta}\right) dv . \qquad (8.18)$$

The postulate of irreversibility, also called the second law of thermodynamics states that

$$\dot{H} - \oint_S \frac{q^k ds_k}{\theta} - \int_v \frac{\varrho h}{\theta} dv \geqslant 0 . \qquad (8.19)$$

In the form (8.19) this law is also known as the Clausius-Duhem inequality, or the entropy inequality. In the local form this law reads

$$\varrho\theta\dot{\eta} - \varrho h - q^k_{,k} + \frac{1}{\theta}\theta_{,k} q^k \geqslant 0 . \qquad (8.20)$$

Sometimes it is convenient to use the Helmholtz free energy ψ per unit mass, defined by the relation

$$\psi = \varepsilon - \theta\eta . \qquad (8.21)$$

In substituting this equation into (8.14) we find

$$\varrho\dot{\psi} + \varrho\eta\dot{\theta} = {}_E w . \qquad (8.22)$$

Using (8.11) we may rewrite (8.20) in the form which includes the mechanical working w,

$$-\varrho\dot{\varepsilon} + \varrho\theta\dot{\eta} + w + \frac{1}{\theta}\theta_{,k}q^k \geq 0 , \tag{8.23}$$

and if we introduce the free energy into this inequality, it becomes

$$-\varrho\dot{\psi} - \varrho\eta\dot{\theta} + w + \frac{1}{\theta}\theta_{,k}q^k \geq 0 . \tag{8.24}$$

A process in which

$\dot{\theta} = 0$ is called <u>isothermal</u> ,
$Q = 0$ is called <u>adiabatic</u> ,
$\dot{\eta} = 0$ is called <u>isentropic</u> ,
$\dot{\varepsilon} = 0$ is called <u>isoenergetic</u> .

When in (8.19) we have the equality, we have the case of <u>equilibrium</u> and the corresponding process is <u>reversible</u>.

From (8.14) and (8.22) we see that the strain energy σ is equal to the internal energy ε if the process is isentropic, and the strain energy σ is equal to the free energy ψ if the process is isothermal.

An inspection of (8.8) shows that for the recoverable part of working we may write

$$\begin{aligned} {}_E w &= {}_E t^{(ij)} d_{ij} + {}_E h^{(\lambda)jk} \dot{d}^{i}_{(\lambda),k} w_{ij} + \\ &\quad + {}_E h^{(\lambda)k}_{j} \dot{d}^{j}_{(\lambda),k} - {}_E m^{ijk} w_{ij,k} . \end{aligned} \tag{8.25}$$

Since

$$
{}_Et^{(ij)}d_{ij} = g_{il}\,{}_Et^{(ij)}v^l_{,j} ,
$$

$$
{}_Eh^{(\lambda)jk}d^i_{(\lambda),j}w_{ij} = g_{il}d^{[i}_{(\lambda),k}\,{}_Eh^{(\lambda)j]k}v^l_{,j} , \tag{8.26}
$$

$$
{}_Em^{ijk}w_{ij,k} = g_{il}\,{}_Em^{ijk}v^l_{,jk} = g_{il}\,{}_Em^{i(jk)}v^l_{,jk}
$$

and since

$$
v^l_{,j} = \overline{\dot{x}^l_{;L}}X^L_{;j} ,
$$

$$
v^l_{,jk} = (v^l_{,j})_{,k} = (\overline{\dot{x}^l_{;L}}X^L_{;j})_{,k} = \overline{\dot{x}^l_{;LK}}X^L_{;j}X^K_{;k} + \overline{\dot{x}^l_{;L}}X^L_{;jk} , \tag{8.27}
$$

$$
\dot{d}^j_{(\lambda),k} = \overline{\dot{d}^j_{(\lambda);K}}X^K_{;k} ,
$$

we see that ${}_Ew$ may be expressed as a linear function in the material derivatives of the gradients of deformation and of the directors,

$$
\overline{\dot{x}^l_{;L}} , \quad \overline{\dot{x}^l_{;KL}} , \quad \overline{\dot{d}^j_{(\lambda);K}} .
$$

Thus,

$$
{}_Ew = g_{il}[{}_Et^{(ij)}X^L_{;j} + d^{[i}_{(\lambda),k}\,{}_Eh^{(\lambda)j]k}X^L_{;j} - {}_Em^{i(jk)}X^L_{;jk}]\overline{\dot{x}^l_{;L}} +
$$
$$
+ {}_Eh^{(\lambda)\cdot k}_{\;\;j}X^K_{;k}\overline{\dot{d}^j_{(\lambda);K}} - g_{il}\,{}_Em^{i(jk)}X^L_{;j}X^K_{;k}\overline{\dot{x}^l_{;KL}} . \tag{8.28}
$$

According to (8.14), we may assume that the internal energy is a function of the deformation and director gradients and of the entropy,

$$\varepsilon = \varepsilon(x^{l}_{;L}\, , \, x^{l}_{;KL}\, , \, d^{l}_{(\lambda);K}\, , \, \eta)$$

so that*

$$(8.29) \qquad \dot{\varepsilon} = \frac{\partial \varepsilon}{\partial x^{l}_{;L}} \dot{\overline{x^{l}_{;L}}} + \frac{\partial \varepsilon}{\partial x^{l}_{;KL}} \dot{\overline{x^{l}_{;KL}}} + \frac{\partial \varepsilon}{\partial d^{l}_{(\lambda);K}} \dot{\overline{d^{l}_{(\lambda);K}}} + \frac{\partial \varepsilon}{\partial \eta} \dot{\eta} .$$

Since the relation (8.14) must be valid for any processes, it must be satisfied for arbitrary rates $\dot{\overline{x^{l}_{;L}}}$, $\dot{\overline{x^{l}_{;KL}}}$, $\dot{\overline{d^{l}_{(\lambda);K}}}$ and $\dot{\eta}$, which yields the following relations

$$(8.30) \qquad {}_{E}m^{i(jk)} = -\varrho g^{il} \frac{\partial \varepsilon}{\partial x^{l}_{;KL}} x^{j}_{;K} x^{k}_{;L} ,$$

$$(8.31) \qquad {}_{E}h^{(\lambda)jk} = \varrho g^{jl} \frac{\partial \varepsilon}{\partial d^{l}_{(\lambda);K}} x^{k}_{;K} ,$$

$$(8.32) \quad {}_{E}t^{(ij)} = \varrho\left[g^{il}\left(\frac{\partial \varepsilon}{\partial x^{l}_{;L}} x^{j}_{;L} + \frac{\partial \varepsilon}{\partial x^{l}_{;KL}} x^{j}_{;KL} \right) + \left(g^{il} \frac{\partial \varepsilon}{\partial d^{l}_{(\lambda);K}} d^{j}_{(\lambda);k} \right)_{[ij]} x^{k}_{;K} \right] .$$

Hence, from the first law of thermodynamics we may obtain certain relations for the reversible parts of the symmetric part of the stress tensor, and of the symmetric part of the couple-stress tensor and for the director-stress tensor. The dissipative parts

---

* We follow here the procedure applied by Stojanović and Djurić [425, 427] and by Stojanović, Djurić and Vujosević [343] in the case of elasticity.

remain unchanged.

Regarding the dissipative parts of the stress tensor, couple-stress tensor and of the director stress tensors, there is a discussion whether or not the inequalities (8.19), or (8.23, 24) present any restrictions. E.g. Kline [235] demonstrated that from these inequalities without additional assumptions further conclusions cannot be made, but Leigh [265] (in the non-polar case) finds certain restrictions and applies the second law of thermodynamics to plasticity and linear viscous flow. Green and Rivlin [176] obtained the differential equations of theories of generalized continua by the systematic use of the first and second law of thermodynamics, but applied the procedure only to the reversible case (cf. also Green and Laws [154]).

I find, however, that in some cases the principle of least irreversible force by Ziegler [516] is very useful.* Ziegler applied it to a number of cases in the theory of non-polar materials.

For polar materials this principle was applied for the derivation of the constitutive relations of plasticity and viscous flow by Komljenović [243], Plavsić [359, 361], Plavsić and Stojanović [363] and Djurić [88].

Ziegler assumed that the entropy η has two parts,

---

* This principle is not generally accepted and some authors have serious objections on its general validity.

the irreversible part $\eta^{(i)}$ and the irreversible part $\eta^{(r)}$, so that

(8.33) $$\eta = \eta^{(i)} + \eta^{(r)} ,$$

and

(8.34) $$\varrho\theta\dot{\eta}^{(i)} = {}_D w ,$$
$$\varrho\theta\dot{\eta}^{(r)} = q^k_{,k} + \varrho h .$$

These relations satisfy the equation of production of entropy. Further he assumed the second law of thermodynamics (for $dt > 0$) to be of the form

(8.35) $$\dot{\eta}^{(i)} \geqslant 0 .$$

From (8.20) we see that this assumption is valid only if

(8.36) $$\theta_{,k} q^k \leqslant 0 ,$$

which is not in contradiction with the experience, since the temperature flows from the parts of the body with higher temper ature to the parts with lower temperature. It follows then from (8.34) that

(8.37) $${}_D w \geqslant 0 .$$

The rate of entropy production $\dot{\eta}^{(i)}$ is independent of the heat exchange and may be a function of the rates of deformation only.

If x^k, $k = 1,..,n$ are variables which describe the configuration of a thermodynamical system and if $X_k^{(i)}$ are irreversible forces, we may write

$$ {}_D w = X_k^{(i)} dx^k . \tag{8.38}$$

In an n-dimensional space of the variables x^k the <u>dissipation function</u>

$$\Phi(\dot{\underset{\sim}{x}}) = \theta\dot{\eta}^{(i)} \tag{8.39}$$

for each prescribed value of the velocities $\dot{x}^k$ represents a surface,

$$\Phi(\dot{\underset{\sim}{x}}) = M . \tag{8.40}$$

Assuming that a process considered is quasistatic, i.e. the change of the coordinates x^k and of the temperature θ is sufficiently slow, the <u>principle of least irreversible force</u> states that:

If the value $M > 0$ of the dissipation function $\Phi(\dot{x}^k)$ and the direction ν_k of the irreversible force $(X_k^{(i)} = X\nu_k)$ are prescribed, then the actual quasistatic velocity $\dot{x}^k$ minimizes the magnitude X of the irreversible force $X_k^{(i)}$ subject to the condition $\Phi(\dot{x}^k) \geq 0$.

For the justification of this principle we refer to Ziegler's paper [516] .

As a consequence of this principle it follows

that the components of the irreversible force have to satisfy the equations

$$X_k^{(i)} = \lambda \frac{\partial \Phi}{\partial \dot{x}^k} , \tag{8.41}$$

where

$$\lambda = \Phi \left(\frac{\partial \Phi}{\partial \dot{x}^m} \dot{x}^m \right)^{-1} . \tag{8.42}$$

When we identify the components $X_k^{(i)}$ with the components of the irreversible parts of the stress tensor, couple stress tensor and tensor of the director stresses, and the velocities $\dot{x}^k$ with the corresponding rates of the deformation of position and directors, from (8.40) follow the relations for ${}_D\underset{\sim}{t}$, ${}_D\underset{\sim}{m}$ and ${}_D\underset{\sim}{h}^{(\lambda)}$.

8.1 Invariance of the First Law of Thermodynamics and the Equations of Motion

The first law of thermodynamics may be written in the explicit form (cf. 8.10)

$$\begin{aligned} \frac{d}{dt}\int_v \varrho\left(\varepsilon + \frac{1}{2}\underset{\sim}{v}\cdot\underset{\sim}{v} + \frac{1}{2}i^{\lambda\mu}\dot{\underset{\sim}{d}}_{(\lambda)}\cdot\dot{\underset{\sim}{d}}_{(\mu)}\right)dv &= \\ = \oint_s (\underset{\sim}{t}\cdot\underset{\sim}{v} + \underset{\sim}{h}^{(\lambda)}\cdot\dot{\underset{\sim}{d}}_{(\lambda)} - \underset{\sim}{m}\cdot\underset{\sim}{w} + q)ds &+ \\ + \int_v \varrho(\underset{\sim}{f}\cdot\underset{\sim}{v} + \underset{\sim}{k}^{(\lambda)}\cdot\dot{\underset{\sim}{d}}_{(\lambda)} - \underset{\sim}{\ell}\cdot\underset{\sim}{w} + h)dv & . \end{aligned} \tag{8.1.1}$$

where we have put

$$\underset{\sim}{t}\cdot\underset{\sim}{v} = t^{ij}v_i n_j , \quad \underset{\sim}{h}^{(\lambda)}\cdot\dot{\underset{\sim}{d}}_{(\lambda)} = h^{(\lambda)ij}d_{(\lambda)i}n_j ,$$

$$\underset{\sim}{m}\cdot\underset{\sim}{w} = m^{ijk}w_{ij}n_k , \quad \underset{\sim}{k}^{(\lambda)}\cdot\dot{\underset{\sim}{d}}_{(\lambda)} = k^{(\lambda)i}\dot{d}_{(\lambda)i} , \qquad (8.1.2)$$

$$\underset{\sim}{\ell}\cdot\underset{\sim}{w} = \ell^{ij}w_{ij} ,$$

and $\underset{\sim}{w}$ is the antisymmetric vorticity tensor defined by (3.28).

Two motions of a body considered differ by a rigid body motion if the velocities of the points of the body differ by a rigid body velocity. Let $\underset{\sim}{v}$ and $\underset{\sim}{v}^*$ be two velocities of a point $\underset{\sim}{X}$, and let $\underset{\sim}{a}$ and $\underset{\sim}{\omega}$ be two constant vectors. A rigid motion is defined by the velocity field

$$\underset{\sim}{v} = \underset{\sim}{a} + \underset{\sim}{\omega}\times(\underset{\sim}{r} - \underset{\sim}{r}_0) = \underset{\sim}{b} + \underset{\sim}{\omega}\times\underset{\sim}{r} \qquad (8.1.3)$$

where $\underset{\sim}{r}$ is the position of $\underset{\sim}{X}$ and $\underset{\sim}{r}_0$ is an arbitrary constant vector. Since $\underset{\sim}{\omega}$ is the angular velocity vector, its components are $\omega_i = \frac{1}{2}\varepsilon_{ijk}\omega^{jk}$ and $\omega^{jk} = -\omega^{kj}$.

We postulate now the invariance of (8.1.1) under superposed rigid body motions. This means that the form of (8.1.1) is invariant for all motions which differ by an arbitrary rigid motion.

When in (8.1.1) $\underset{\sim}{v}$ is substituted by $\underset{\sim}{v} + \underset{\sim}{b}$, the postulate will be satisfied only if

$$(8.1.4) \qquad \int_{v} \left[\varrho \dot{\underset{\sim}{v}} \cdot \underset{\sim}{b}\, dv + \left(\underset{\sim}{v} \cdot \underset{\sim}{b} + \frac{1}{2} b^2 \right) \dot{\overline{\varrho\, dv}} \right] = \oint_{S} \underset{\sim}{t} \cdot \underset{\sim}{b}\, ds + \int_{v} \varrho \underset{\sim}{f} \cdot \underset{\sim}{b}\, dv \; .$$

For arbitrary $\underset{\sim}{b}$ and $\underset{\sim}{b}^2$ we obtain two relations, the law of conservation of mass,

$$\dot{\overline{\varrho\, dv}} = 0 \; ,$$

which by (3.46) obtains the usual form

$$(8.1.5) \qquad J\varrho = \varrho_0 \; ,$$

and the equation of motion (7.19),

$$(8.1.6) \qquad \varrho \dot{v}^i = t^{ij}{}_{,j} + \varrho f^i \; .$$

To investigate the consequences of the invariance of (8.1.1) under superposed arbitrary rigid body rotations, we have to substitute $\underset{\sim}{v}$, $\dot{\underset{\sim}{v}}$, $\dot{\underset{\sim}{d}}_{(\lambda)}$, $\ddot{\underset{\sim}{d}}_{(\lambda)}$ and $\underset{\sim}{w}$ by

$$\underset{\sim}{v} \rightarrow \underset{\sim}{v} + \underset{\sim}{\omega} \times \underset{\sim}{r}$$

$$(8.1.7) \qquad \dot{\underset{\sim}{v}} \rightarrow \dot{\underset{\sim}{v}} + \underset{\sim}{\omega} \times (\underset{\sim}{v} + \underset{\sim}{\omega} \times \underset{\sim}{r})$$

$$\dot{\underset{\sim}{d}}_{(\lambda)} \rightarrow \dot{\underset{\sim}{d}}_{(\lambda)} + \underset{\sim}{\omega} \times \underset{\sim}{d}_{(\lambda)}$$

$$\ddot{\underset{\sim}{d}}_{(\lambda)} \rightarrow \ddot{\underset{\sim}{d}}_{(\lambda)} + \underset{\sim}{\omega} \times (\dot{\underset{\sim}{d}}_{(\lambda)} + \underset{\sim}{\omega} \times \underset{\sim}{d}_{(\lambda)}) \; ,$$

respectively. Thus we obtain the relation

$$\int_v \rho\,\underset{\sim}{\omega}\cdot[\underset{\sim}{r}\times\dot{\underset{\sim}{v}} + i^{\lambda\mu}\underset{\sim}{d}_{(\lambda)}\times\ddot{\underset{\sim}{d}}_{(\mu)}]dv =$$

$$= \int_v \rho\,\underset{\sim}{\omega}\cdot(\underset{\sim}{r}\times\underset{\sim}{f} + \underset{\sim}{d}_{(\lambda)}\times\underset{\sim}{k}^{(\lambda)} + \underset{\sim}{\ell})dv + \tag{8.1.8}$$

$$+ \oint_S \underset{\sim}{\omega}\cdot(\underset{\sim}{r}\times\underset{\sim}{t} + \underset{\sim}{d}_{(\lambda)}\times\underset{\sim}{h}^{(\lambda)} + \underset{\sim}{m})ds\,.$$

Using (8.1.6) and after the application of the divergence theorem for arbitrary rotations $\underset{\sim}{\omega}$ we obtain

$$\rho i^{\lambda\mu} d_{(\lambda)}^{[i}\ddot{d}_{(\mu)}^{j]} = t^{[ji]} + (\rho d_{(\lambda)}^{[i}k^{(\lambda)j]} + \ell^{ij}) + (d_{(\lambda)}^{[i}h^{(\lambda)j]k} + m^{ijk})_{,k} \tag{8.1.9}$$

which coincides with (7.34).

The equations of motion of the directors (7.33) may be obtained from the invariance of the relation (8.1.1) under arbitrary rigid translations of the directors. If $\underset{\sim}{C}_{(\lambda)}$ are arbitrary constant vectors, and if we substitute $\underset{\sim}{d}_{(\lambda)}$ by

$$\dot{\underset{\sim}{d}}_{(\lambda)} + \underset{\sim}{C}_{(\lambda)}$$

in (8.1.1), it follows immediately that the form of the first law of thermodynamics will be preserved if

$$C_{(\lambda)i}(\rho i^{\lambda\mu}\ddot{d}_{(\mu)}^{i} - \rho k^{(\lambda)i} - h^{(\lambda)ij}{}_{,j}) = 0\,,$$

which for arbitrary $C_{(\lambda)i}$ reduces to (7.33).

This last requirement, that (8.1.1) is invariant if the rates of the directors are changed by some arbitrary, constant rates, is an extension of the well-known invariance of the energy-balance law under superposed rigid motions in classical continuum mechanics. This extension however, is not unnatural since (8.1.3) are related to the displacements of the points of the body, and the motions of the directors are independent of the motions of the points. That was the principal reason for our introduction of this new, additional requirement for the invariance of (8.1.1).

From the results in this Section we see that the postulated invariance of the first law of thermodynamics under arbitrary rigid motions of the points and of the directors is equivalent with the principles of balance of the Section 7, and contains all these separate principles as special cases.

9. Some General Considerations on Constitutive Relations

The relations (8.30-32) for the reversible part of the stress, director stresses and couple-stress tensor, as well as the relations for the irreversible parts which follow from (8.40), have to satisfy some additional assumptions in order to represent constitutive relations.

Constitutive relations in mechanics describe the response of a material to deformations. The response is characterized by the intrinsic properties of matter and not by the choice of coordinates, or by the choice of the way of describing deformations, rates of deformation, motions etc. Constitutive relations,never describe completely mechanical properties of real materials, but only some of the dominant properties considered for some particular purposes. Therefore, a material which would completely behave according to some prescribed constitutive relations is an ideal material and does not exist in the Nature.

The first question, regarding the constitutive relations, is: which quantities are to be determined by these relations and which quantities are to be considered as variables. There are 3n+3 differential equations (7.35) and (7.36) from which the motions $\underset{\sim}{x} = \underset{\sim}{x}(\underset{\sim}{X},t)$ and $\underset{\sim}{d}_{(\alpha)} = \underset{\sim}{d}_{(\alpha)}(\underset{\sim}{X},t)$ may be determined if the forces $\underset{\sim}{f}$ and $\underset{\sim}{k}^{(\alpha)}$ and the couples $\underset{\sim}{\ell}$ are prescribed, but there are 9+9+27=45 components of the tensors $\underset{\sim}{t}$, $\underset{\sim}{h}^{(\alpha)}$ and $\underset{\sim}{m}$ which cannot be determined from these equations. If we turn to the laws of thermodynamics, we obtain some relations, but then two new additional quantities are introduced, temperature θ and entropy η. Expressing the laws of thermodynamics in terms of the internal energy ε, or in terms of the free energy ψ we may regard θ , or η respectively, as a quantity to be determined by a constitutive relation.

There are two methods for the formulation of constitutive relations. One method is: to assume certain relations and to subject them to certain restrictions which follow from thermodynamics and from the principles which will be introduced later. The other method consists in deriving the relations from the energetic considerations based on thermodynamics; so obtained relations are then to be subjected to further restrictions furnished by the additional principles.

The number of the assumed additional principles which are to be imposed on the constitutive relations varies from author to author. Since we are going to consider the constitutive relations which follow from the energetic considerations, and since we are not going to consider problems of more complex nature such as viscoelasticity and dependence of the state of stress on the history of deformation, we shall restrict the number of additional assumptions to two principles,

1° The principle of material frame indifference, and

2° The principle of local action.

The discussion of various other principles in continuum mechanics may be found e.g. in the books by Truesdell and Noll [468] and by Eringen [122, 131].

The two mentioned principles are independent of the so called material symmetries. In order to obtain the relations for a particular class of material symmetries, we have to require, in addition, that the constitutive relations are invariant with respect to a subgroup of the group of orthogonal transformations which characterizes the class of material symmetries considered.

Let z^{α} and $\bar{z}^{\alpha}$ be two orthogonal Cartesian coordin ate systems with origins at $\underset{\sim}{O}$ and $\underset{\sim}{\bar{O}}$, and let an <u>event</u> be described with respect to these two systems by $\{\underset{\sim}{z}, t\}$ and $\{\underset{\sim}{\bar{z}}, \bar{t}\}$, where t and $\bar{t}$ are times measured by two observers at $\underset{\sim}{O}$ and $\underset{\sim}{\bar{O}}$. A change of the frame of reference is expressed by the formula

$$\bar{z}^{\alpha} = Q^{\alpha}_{\cdot\beta}(t)\, z^{\beta} + a^{\alpha}(t) \qquad (9.1)$$

$$\bar{t} = t - \tau ,$$

or

$$z^{\alpha} = Q^{\cdot\alpha}_{\beta}(\bar{t})\, \bar{z}^{\beta} + b^{\alpha}(\bar{t}) \qquad (9.2)$$

$$t = \bar{t} + \tau .$$

Here

$$Q^{\cdot\alpha}_{\beta} Q^{\cdot\gamma}_{\alpha} = \delta^{\gamma}_{\beta} , \quad Q^{\cdot\alpha}_{\beta} Q^{\cdot\beta}_{\gamma} = \delta^{\alpha}_{\gamma} , \qquad (9.3)$$

and we assume that $\underset{\sim}{Q}$ is an orthogonal matrix, $\underset{\sim}{Q}^{-1} = \underset{\sim}{Q}^{T}$.

If $\underset{\sim}{T}$ is a tensor field with components $T^{\cdots}_{\cdots}$ and $\bar{T}^{\cdots}_{\cdots}$ with respect to the coordinate systems $\underset{\sim}{O}\underset{\sim}{z}$ and $\underset{\sim}{\bar{O}}\underset{\sim}{\bar{z}}$ respectively, and if the components transform according to the transformation law for tensors when both, the dependent and independent variables, are transformed according to (9.1.2), the tensor field $\underset{\sim}{T}$ is said to be <u>frame-indifferent</u>, or <u>objective</u>.

The components of the position vector $\underset{\sim}{r} = z^{\alpha}\underset{\sim}{e}_{\alpha}$ are obviously not objective quantities since they transform according to (9.1.2).

The components of the velocity vector $\underset{\sim}{v}$ are defined with respect to the two considered reference frames by

$$(9.4) \qquad v^{\alpha} = \dot{z}^{\alpha} , \quad \bar{v}^{\alpha} = \dot{\bar{z}}^{\alpha} .$$

From (9.1) we have

$$(9.5) \qquad \bar{v}^{\alpha} = \dot{Q}^{\alpha}_{\cdot\beta} z^{\beta} + Q^{\alpha}_{\cdot\beta} v^{\beta} + \dot{a}^{\alpha}$$

and obviously the velocity vector is not an objective vector. Writing (9.5) in the form

$$(9.6) \qquad \bar{v}_{\alpha} = \dot{Q}_{\alpha\lambda} z^{\lambda} + Q_{\alpha}^{\cdot\lambda} v_{\lambda} + \dot{a}_{\alpha}$$

we obtain for the velocity gradients the following transformation law,

$$(9.7) \qquad \frac{\partial \bar{v}_{\alpha}}{\partial \bar{z}^{\beta}} = \dot{Q}_{\alpha\lambda} \frac{\partial z^{\lambda}}{\partial \bar{z}^{\beta}} + Q_{\alpha}^{\cdot\lambda} \frac{\partial v_{\lambda}}{\partial z^{\mu}} \frac{\partial z^{\mu}}{\partial \bar{z}^{\beta}} = \dot{Q}_{\alpha\lambda} Q_{\beta}^{\cdot\lambda} + Q_{\alpha}^{\cdot\lambda} \frac{\partial v_{\lambda}}{\partial z^{\mu}} Q_{\beta}^{\cdot\mu} .$$

Hence, the velocity gradients are not objective quantities. However, the rate of strain is an objective tensor. From (9.7) we have

$$\bar{d}_{\alpha\beta} = \frac{1}{2}\left(\frac{\partial \bar{v}_{\alpha}}{\partial \bar{z}^{\beta}} + \frac{\partial \bar{v}_{\beta}}{\partial \bar{z}^{\alpha}}\right) = Q_{(\beta}^{\cdot\lambda} \dot{Q}_{\alpha)\lambda} + Q_{\alpha}^{\cdot\lambda} Q_{\beta}^{\cdot\mu} v_{(\lambda,\mu)} ,$$

but in view of (9.3)

$$Q_{\beta}^{\cdot\lambda}\dot{Q}_{\alpha\lambda} + Q_{\alpha}^{\cdot\lambda}\dot{Q}_{\beta\lambda} = \frac{d}{dt}(Q_{\beta}^{\cdot\lambda}Q_{\alpha\lambda}) = 0 ,$$

and obviously

$$\bar{d}_{\alpha\beta} = Q_{\alpha}^{\cdot\lambda}Q_{\beta}^{\cdot\mu}d_{\lambda\mu} .$$

From (9.7) it may be seen that the vorticity tensor $w_{\alpha\beta} = v_{[\alpha,\beta]}$ is not an objective tensor, but the gradients of this tensor are objective quantities. We have

$$w_{\alpha\beta} = Q_{[\beta}^{\cdot\lambda}\dot{Q}_{\alpha]\lambda} + Q_{\alpha}^{\cdot\lambda}Q_{\beta}^{\cdot\mu}w_{\lambda\mu}$$

and

$$\bar{w}_{\alpha\beta,\gamma} = Q_{\alpha}^{\cdot\lambda}Q_{\beta}^{\cdot\mu}Q_{\gamma}^{\cdot\nu}w_{\lambda\mu,\nu} . \qquad (9.8)$$

If points of a body are referred to a system of material Cartesian coordinates Z^{λ} and if z^{α} and $\bar{z}^{\alpha}$ are two spatial reference frames, we see from (9.1) that

$$\frac{\partial \bar{z}^{\alpha}}{\partial Z^{\lambda}} = Q^{\alpha}_{\cdot\beta}\frac{\partial z^{\beta}}{\partial Z^{\lambda}} , \qquad (9.9)$$

and the deformation gradients are objective. The same holds for the higher order deformation gradients

$$\frac{\partial^2 \bar{z}^{\alpha}}{\partial Z^{\lambda}\partial Z^{\mu}} = Q^{\alpha}_{\cdot\mu}\frac{\partial^2 z^{\mu}}{\partial Z^{\lambda}\partial Z^{\mu}} , \quad \text{etc.} \qquad (9.10)$$

The <u>principle of material frame indifference</u> requires that: Constitutive equations must be invariant with respect to rigid motions of the spatial frame of reference.

A function $F(V^{\alpha}_{(1)}, V^{\alpha}_{(2)}, \ldots, z^{\alpha})$ of vectors $\underset{\sim}{V}_{(\lambda)}$ is objective or frame-indifferent if it remains invariant under rigid motions of the spatial frame.

If only translations are regarded, $\bar{z}^{\alpha} = z^{\alpha} + a^{\alpha}$, it follows that

$$\bar{V}^{\alpha}_{(\lambda)} = V^{\alpha}_{(\lambda)}$$

and the condition of objectivity for the function F reduces to

$$F(\underset{\sim}{V}_{(1)}, \ldots, z^{\alpha} + a^{\alpha}) = F(\underset{\sim}{V}_{(1)}, \ldots, z^{\alpha}) .$$

If the translations a^{α} are small quantities, from the Taylor series expansion of the function F we obtain that it will be objective only if

$$\frac{\partial F}{\partial z^{\alpha}} = 0$$

i.e. if it does not depend explicitly on spatial coordinates of position.

Let us see now which restrictions are imposed on the function F by arbitrary rigid rotations of the spatial frame, if F is an objective function.

Let Q be the matrix

$$\underset{\sim}{Q} = (\delta^{\alpha}_{\beta} + \omega^{\alpha}_{\cdot\beta}) ,$$

where $\omega_{\alpha\beta} = -\omega_{\beta\alpha}$ is an arbitrary infinitesimal rotation, and

$$\bar{z}^{\alpha} = Q^{\alpha}_{.\beta} z^{\beta} = (\delta^{\alpha}_{\beta} + \omega^{\alpha}_{.\beta}) z^{\beta} . \tag{9.11}$$

If F is an objective function of vectors $V^{\alpha}_{(\nu)}$, $\nu = 1,2,..,n$, it will satisfy the relation

$$F(V^{\alpha}_{(1)}, \ldots, V^{\alpha}_{(n)}) = \bar{F}(\bar{V}^{\alpha}_{(1)}, \ldots, \bar{V}^{\alpha}_{(n)}) . \tag{9.12}$$

From (9.11) we have

$$\bar{V}^{\alpha}_{(\nu)} = V^{\beta}_{(\nu)} \frac{\partial \bar{z}^{\alpha}}{\partial z^{\beta}} = V^{\alpha}_{(\nu)} + V^{\beta}_{(\nu)} \omega^{\alpha}_{.\beta} \tag{9.13}$$

and the invariance requirement (9.12) reduces to the relation

$$F(V^{\alpha}_{(1)}, \ldots) = F(V^{\alpha}_{(1)} + V^{\beta}_{(1)} \omega^{\alpha}_{.\beta}, \ldots) . \tag{9.14}$$

For sufficiently small $\omega^{\alpha}_{.\beta}$ we may expand F into the Taylor series,

$$F(V^{\alpha}_{(1)} + V^{\beta}_{(1)} \omega^{\alpha}_{.\beta}, \ldots) = F(V^{\alpha}_{(1)}, \ldots) + \sum_{\nu=1}^{n} \frac{\partial F}{\partial V^{\alpha}_{(\nu)}} V^{\beta}_{(\nu)} \omega^{\alpha}_{.\beta} + \ldots, .$$

Hence, if F is an objective function, for infinitesimal rotations $\underset{\sim}{\omega}$ we obtain that F has to satisfy the condition

$$\sum_{\nu=1}^{n} \frac{\partial F}{\partial V^{\alpha}_{(\nu)}} V^{\beta}_{(\nu)} \omega^{\alpha}_{.\beta} = 0 . \tag{9.15}$$

But ω is an arbitrary antisymmetric tensor and (9.15) reduces to the system of three differential equations (Toupin [460])

$$\left(\sum_{\nu=1}^{n} \frac{\partial F}{\partial V^{\alpha}_{(\nu)}} V_{(\nu)\beta} \right)_{[\alpha\beta]} = 0 . \tag{9.16}$$

The equations of (9.16) are tensorial equations. If the variables are objective quantities, we may write (9.16) in the form appropriate for arbitrary curvilinear coordinates x^k,

$$\left(\sum_{\nu=1}^{n} g^{il} \frac{\partial F}{\partial V^{l}_{(\alpha)}} V^{j}_{(\alpha)}\right)_{[ij]} = 0 . \tag{9.17}$$

The principle of local action states that: the state of stress at a point $\underset{\sim}{Z}$ of a medium is determined by the motion inside an arbitrary neighborhood $N(\underset{\sim}{Z})$ of the point $\underset{\sim}{Z}$, and the motion outside this neighbourhood may be disregarded.

Under the "state of stress" we understand the values of all the quantities which describe the stress field ($\underset{\sim}{t}$, $\overset{(\lambda)}{\underset{\sim}{h}}$, $\underset{\sim}{m}$ etc.). If $\varphi(z(Z))$ is a function which describes the state of stress at Z at time t, according to this principle, at a configuration $K(t)$ the state of stress at Z is determined by the instantaneous configuration of the neighbourhood $N(\underset{\sim}{Z})$. Let $\underset{\sim}{Z}'$ be a point in $N(\underset{\sim}{Z})$. At the configuration $K(t)$ the relative position of $\underset{\sim}{Z}'$ with respect to $\underset{\sim}{Z}$ is given by

$$\Delta \underset{\sim}{z} = \underset{\sim}{z}(\underset{\sim}{Z}', t) - \underset{\sim}{z}(\underset{\sim}{Z}, t) .$$

If $Z'^{\alpha} - Z^{\alpha} = \Delta Z^{\alpha}$, we may write

$$\Delta z^{\alpha} = \frac{\partial z^{\alpha}}{\partial Z^{\lambda}} \Delta Z^{\lambda} + \frac{1}{2} \frac{\partial^2 z^{\alpha}}{\partial Z^{\lambda} \partial Z^{\mu}} \Delta Z^{\lambda} \Delta Z^{\mu} + \dots . \tag{9.18}$$

Since the state of stress at $\underset{\sim}{Z}$ is determined by the local configuration of an arbitrary neighbourhood $N(\underset{\sim}{Z})$, it follows that

ψ must be a function of the deformation gradients,

$$\psi = \psi(z^{\alpha}_{;\lambda_1}, z^{\alpha}_{;\lambda_1\lambda_2}, \dots z^{\alpha}_{;\lambda_1\dots\lambda_N}, \dots z, t) . \tag{9.19}$$

If ψ is the internal energy function $\mathcal{E}$ and if N is the highest order of the deformation gradients which appears in the expression for the energy, according to Toupin [463], the corresponding material is said to be of <u>order of N.</u>

Stojanović and Djurić [425, 426] generalized this notion to directed elastic bodies, considering the strain energy as a function of the deformation gradients of an order N, and of the director gradients of an order M, such that $\mathcal{E}$ is a function of the form *

$$\mathcal{E} = \mathcal{E}(\overset{k}{x}_{;K}, \overset{k}{x}_{;K_1K_2}, \dots, \overset{k}{x}_{;K_1\dots K_N}; \overset{k}{d}_{(\lambda);K}, \overset{k}{d}_{(\lambda);K_1K_2}, \overset{k}{d}_{(\lambda)K_1K_2\dots K_M}; \eta, \underset{\sim}{X}) . \tag{9.20}$$

In the following we restrict our considerations to the materials of the order $N=2$ and $M=1$, i.e. the constitutive variables, which are to be considered as independent variables, in the expression for the internal energy density are first and

---

* A number of authors considered the strain energy as a function of the components $\overset{k}{d}_{(\lambda)}$ of directors, and not only as a function of the gradients of the directors (mostly in linear theories). From our considerations in the section 8 (see eq. (8.28)) it does not follow that the components of the directors appear explicitly as constitutive variables and therefore we omit them here.

second order deformation gradients and the director gradients, so that

$$\varepsilon = \varepsilon(x^k_{;K}\, x^k_{;KL}\, , d^k_{(\lambda);K}\, , \eta\, , \underset{\sim}{X}) \; . \tag{9.21}$$

Generalizations to higher order materials are in principle simple, but require more involved notation which makes the expression less clear. The higher order gradients of deformation and directors may be identified with the multipolar displacements, and the theory then might be directly applied.

The materials for which the constitutive relations do not depend explicitly on $\underset{\sim}{X}$ are called homogeneous and we shall consider only such materials.

9.1 The Internal Energy Function

The internal energy function ε in the form (9.21) has, according to the principle of material frame indifference, to satisfy the conditions of the form (9.17). When the constitutive variables are identified with the components of the vectors $V_{(\alpha)}$ according to the table

$$V^l_{(1)}\, , V^l_{(2)}\, , V^l_{(3)} \;\rightarrow\; x^l_{;1}\, , x^l_{;2}\, , x^l_{;3}$$

$$V^l_{(4)}\, , \ldots , V^l_{(9)} \;\rightarrow\; x^l_{;11}, \ldots , x^l_{;33}$$

$$V^l_{(10)}, \ldots , V^l_{(3n+9)} \rightarrow d^l_{(1);1}, \ldots , d^l_{(n);3} \; ,$$

the equations (9.17) obtain the form

$$\left[g^{i\ell}\left(\frac{\partial\mathcal{E}}{\partial x^{\ell}_{;K}}x^{j}_{;K}+\frac{\partial\mathcal{E}}{\partial x^{\ell}_{;KL}}x^{j}_{;KL}+\frac{\partial\mathcal{E}}{\partial d^{\ell}_{(\lambda);K}}d^{j}_{(\lambda);K}\right)\right]_{[ij]} = 0 . \qquad (9.1.1)$$

This represents a system of 3 linear partial differential equations with 3 x (3n+9) variables $V^{\ell}_{(\nu)}$, $\ell = 1,2,3$; $\nu = 1,2,\ldots,3n+9$.

The internal energy $\mathcal{E}$ is an arbitrary function of $3\times(3n+9) - 3 = 9n+24$ independent integrals of the system (9.1.1).

It is a matter of a direct calculation to verify that the integrals of the system (9.1.1) are the material tensors

$$C_{AB} \equiv g_{ab}x^{a}_{;A}x^{b}_{;B} , \qquad (9.1.2)$$

$$G_{CAB} = g_{ab}x^{a}_{;CA}x^{b}_{;B} , \qquad (9.1.3)$$

$$F_{\alpha AB} = g_{ab}x^{a}_{;A}d^{b}_{(\alpha);B} . \qquad (9.1.4)$$

These tensors are invariant under the transformations of spatial coordinates. Since

$$C_{AB} = C_{BA} \qquad G_{CAB} = G_{ACB} \qquad (9.1.5)$$

there are $6+18+9n$ independent integrals $\underset{\sim}{C}$, $\underset{\sim}{G}$, $\underset{\sim}{F}_{(\alpha)}$ and the internal energy is an arbitrary function of these quantities,

$$\mathcal{E} = \mathcal{E}(C_{AB}, G_{CAB}, F_{\alpha AB}, X^{K}) . \qquad (9.1.6)$$

9.2 Irreversible Processes

The dissipation function Φ in (8.39) is a function of certain generalized velocities. According to the principle of material frame indifference Φ has to be a function of objective variables. Such variables are the components of the rate of strain tensor $d_{ij} = v_{(i,j)}$, the gradients of vorticity $w_{ij,k}$, as well as the second gradients $v_{i,jk}$ of the velocity vector.

For oriented media the rates of directors $\dot{d}^{i}_{(\alpha)}$ and the gradients $\dot{d}^{i}_{(\alpha);k}$ of these rates are objective tensors. With respect to rigid motions (9.1.3) of Cartesian frames, it follows that the directors are objective vectors,

$$\bar{d}^{\lambda}_{(\alpha)} = d^{\mu}_{(\alpha)} Q^{\lambda}_{\cdot\mu}$$

but the rates

$$\dot{\bar{d}}^{\lambda}_{(\alpha)} = \dot{d}^{\mu}_{(\alpha)} Q^{\lambda}_{\cdot\mu} + d^{\mu}_{(\alpha)} \dot{Q}^{\lambda}_{\cdot\mu}$$

and the gradients of the rates

$$\dot{\bar{d}}^{\lambda}_{(\alpha),\beta} = \dot{d}^{\mu}_{(\alpha),\nu} Q^{\lambda}_{\cdot\mu} Q^{\cdot\nu}_{\beta} + d^{\mu}_{(\alpha),\nu} \dot{Q}^{\lambda}_{\cdot\mu} Q^{\cdot\nu}_{\beta}$$

are obviously not objective quantities.

From (8.8) we have for the dissipative part ${}_{D}w$ of the mechanical power the expression

$$_D w = {}_D t^{(ij)} d_{ij} + {}_D h^{(\lambda)jk} (\dot{d}_{(\lambda)j,k} - w_j^{\cdot i} d_{(\lambda)i,k}) - {}_D m^{ijk} w_{ij,k} . \quad (9.2.1)$$

However, we may write

$$\dot{d}_{(\lambda)j,k} - w_j^{\cdot i} d_{(\lambda)i,k} = (\dot{d}_{(\lambda)j} - w_j^{\cdot i} d_{(\lambda)i})_{,k} + w_{j,k}^{\cdot i} d_{(\lambda)i} , \quad (9.2.2)$$

where

$$\hat{d}_{(\lambda)j} = \dot{d}_{(\lambda)j} - w_j^{\cdot i} d_{(\lambda)i} \quad (9.2.3)$$

is the co-rotational time flux (cf. [469]) of the vector $\underset{\sim}{d}_{(\lambda)}$. It may directly be verified that $\hat{d}_{(\lambda)j}$ is an objective vector. Hence, we may rewrite now (9.2.1) in the form

$$_D w = {}_D t^{(ij)} d_{ij} + {}_D h^{(\lambda)jk} \hat{d}_{(\lambda)j,k} - (m^{ijk} + d^{i}_{(\lambda)} h^{(\lambda)jk}) w_{ij,k} . \quad (9.2.4)$$

Hence, all rates which appear here,

$$d_{ij} , \quad w_{ij,k} , \quad \hat{d}_{(\lambda)j,k} \quad (9.2.5)$$

are objective. It would be natural to assume now that the dissipative function Φ depends on the objective rates (9.2.5). But, according to the definition, Φ is a function of velocities, and therefore it might be regarded as a function of $\dot{x}^k_{;K}$, $\dot{x}^k_{;KL}$, $\dot{d}^{k}_{(\lambda);K}$ via the objective variables (9.2.5).

For the derivation of the constitutive relations for irreversible processes we may turn now to Ziegler's principle, or to consider the Clausius-Duhem inequality. Ziegler's principle of least irreversible force is so far applied only to the case of non-orien ted polar media, where it was assumed (for references see section 8) that

$$\Phi = \Phi(d_{ij}, w_{ij,k}) \ . \tag{9.2.6}$$

Formal difficulties for the application of the Clausius-Duhem inequality are evident, since the internal energy function ε, or the free energy ψ, have to be regarded as functions of $x^k_{;K}$, $x^k_{;KL}$, $d^k_{(\lambda);K}$, and not of the rates (9.2.5). Therefore we may only quote Rivlin [377], who said that "The application of the Clausius-Duhem inequality to inelastic materials is...... questionable. It should, however, be realized that the results obtained from many applications are, in the main, not very strong".

The only possibility which remains is to introduce the constitutive relations by assumption, and in the form which will not violate the laws of motion and the laws of thermodynamics. The form of the assumed relations depends on the mechanical properties which are to be considered. Often in the applications of this method is used the <u>principle of equipresence</u>: A quantity present as an independent variable in one constitutive equation should be also present in all, unless its presence contradicts the laws of physics, or the rules of invariance (cf. [468]). It should be noted that this principle is not generally accepted.

In general, constitutive equations have to be in

accordance with the laws of thermodynamics, i.e. not to violate them. Let us write Clausius-Duhem inequality in the form (8.24),

$$-\varrho\dot{\Psi} - \varrho\eta\dot{\theta} + w + \frac{1}{\theta}\theta_{,k}q^k \geq 0 ,$$

and let us assume that some generalized forces X_k are functions of some generalized velocities v^k, $(k = 1,2,\dots,n)$, of $\varrho, \theta, \varrho_{,i}, \theta_{,i}$, etc. We have

$$w = X_k v^k , \tag{9.2.7}$$

and if we assume the principle of equipresence, the quantities Ψ , η , X_k , q^k have all to be functions of the same set of variables,

$$(\Psi,\eta,X_k,q^k) = \text{fonct.}(\varrho,\varrho_{,i},\theta,\theta_{,i},v^k) . \tag{9.2.8}$$

Introducing this into the Clausius-Duhem inequality we obtain

$$\begin{aligned} -\varrho\left(\frac{\partial\Psi}{\partial\varrho}\dot{\varrho} + \frac{\partial\Psi}{\partial\varrho_{,i}}\dot{\overline{\varrho_{,i}}} + \frac{\partial\Psi}{\partial\theta}\dot{\theta} + \frac{\partial\Psi}{\partial\theta_{,i}}\dot{\overline{\theta_{,i}}} + \frac{\partial\Psi}{\partial v^k}\dot{v}^k\right) - \\ - \varrho\eta\theta + X_k v^k + \frac{1}{\theta}\theta_{,k}q^k \geq 0 . \end{aligned} \tag{9.2.9}$$

According to the law of conservation of mass we have

$$\dot{\varrho} = -\varrho I_d , \tag{9.2.10a}$$

(9.2.10b) $$\dot{\overline{\varrho_{,i}}} = -\varrho_{,i} I_d - \varrho_{,k} \dot{x}^k_{,i} - \varrho \partial_i I_d \, ,$$

where I_d is the first invariant of the rate of strain tensor, and x^k are spatial (three-dimensional) coordinates of position of the points of the medium.

The inequality (9.2.9) has to be satisfied for arbitrary rates $\dot{\theta}$, $\dot{\overline{\theta_{,i}}}$, $\dot{v}^k$ and it follows that the necessary condition for this is that

(9.2.11) $$\eta = -\frac{\partial \Psi}{\partial \theta} \, , \quad \frac{\partial \Psi}{\partial \theta_{,i}} = 0 \, , \quad \frac{\partial \Psi}{\partial v^k} = 0 \, .$$

Thus, the free energy function for irreversible processes reduces to

(9.2.12) $$\Psi = \Psi(\varrho \, , \varrho_{,i} \, , \theta)$$

and the inequality (9.2.9) reduces to

(9.2.13) $$\left(\varrho^2 \frac{\partial \Psi}{\partial \varrho} + \frac{\partial \Psi}{\partial \varrho_{,i}} \varrho_{,i}\right) I_d + \frac{\partial \Psi}{\partial \varrho_{,i}} (\varrho_{,k} \dot{x}^k_{,i} + \varrho \partial_i I_d) + \\ + X_k v^k + \frac{\theta_{,k} q^k}{\theta} \geq 0 \, .$$

Obviously, from this inequality it does not seem possible to derive the constitutive equations, but whatever are the assumed constitutive relations, they have to satisfy the inequality (9.2.13).

In this discussion of the Clausius-Duhem inequality we restricted our considerations to the first gradients of $\mathbf{g}$ and θ, but the procedure might be applied to any grade of the gradients and to any number of the other constitutive variables assumed in the theory.

In the theory of inelastic properties of non-polar media, owing to the recent developments of the thermodynamics of continua, some progress is made by Leigh [265] and Dillon [84].

In the following sections we shall discuss the constitutive relations of some particular media, when the constitutive relations are expressed in the form of functions. More general theories, based on functionals, are not very much developed*.

10. Elasticity

In some modern treatments the difference is made between <u>elastic</u> and <u>hyperelastic</u> materials. Hyperelastic materials are those for which an elastic potential exists and the stresses may be derived from this potential. For elastic materials the existence of such a potential is not necessary. Hyperelastic materials are elastic, but elastic materials are not nec

---

* For some aspects of viscoelasticity we refer the readers to the papers by DeSilva and Kline [83] and by Eringen [123, 130].

essarily hyperelastic. We restrict our considerations, according to this division, to hyperelastic materials.

In the sense of thermodynamics the mechanical work done by a deformation of an elastic material is reversible and it is accumulated in the elastic potential energy σ, so that from (8.12, 13) we have

$$w = {}_E w \,, \quad \sigma = \Sigma \,. \tag{10.1}$$

The local law of balance of energy (8.11) may be written in one of the forms corresponding to (8.14) or (8.22),

$$\varrho \dot{\varepsilon} = {}_E w + \varrho \theta \dot{\eta} \,, \tag{10.2}$$

or

$$\varrho \dot{\Psi} = {}_E w - \varrho \eta \dot{\theta} \,. \tag{10.3}$$

Since the dissipative part of working vanishes we shall drop the subscript "E".

According to the section 8, we assume the specific internal energy to be a function of the form

$$\varepsilon = \varepsilon(x^l_{;L} \,, x^l_{;LK} \,, d^l_{(\lambda);K} \,, \eta) \tag{10.4}$$

and the specific free energy to be a function of the form

$$\Psi = \Psi(x^l_{;L} \,, x^l_{;LK} \,, d^l_{(\lambda);K} \,, \theta) \,. \tag{10.5}$$

If we take the energy balance equation in the

form (10.2), from (8.29, 32) we obtain the following expressions for the temperature, stress, director stress and couple-stress:

$$\theta = \frac{\partial \varepsilon}{\partial \eta} , \tag{10.6}$$

$$t^{(ij)} = \varrho\left[g^{il}\left(\frac{\partial \varepsilon}{\partial x^{l}_{;L}} x^{j}_{;L} + \frac{\partial \varepsilon}{\partial x^{l}_{;KL}} x^{j}_{;KL}\right) + \left(g^{il}\frac{\partial \varepsilon}{\partial d^{l}_{(\lambda);K}} d^{j}_{(\lambda);k}\right)_{[ij]} x^{k}_{;K}\right] , \tag{10.7}$$

$$m^{i(jk)} = -\varrho g^{il} \frac{\partial \varepsilon}{\partial x^{l}_{;KL}} x^{j}_{;K} x^{k}_{;L} , \tag{10.8}$$

$$h^{(\lambda)ij} = \varrho g^{il} \frac{\partial \varepsilon}{\partial d^{l}_{(\lambda);K}} x^{j}_{;K} . \tag{10.9}$$

The similar set of equations follows if the free energy function Ψ is used instead of ε , but since in Ψ the temperature θ is regarded as one of the constitutive variables, the corresponding constitutive equation for entropy will be

$$\eta = -\frac{\partial \Psi}{\partial \theta} . \tag{10.10}$$

The relations (10.7 - 9) cannot be regarded yet as constitutive relations. First, the internal energy must be an objective function, and second, the symmetry properties of the left and right-hand sides of the relations (10.7, 8) have to be the same, i.e. the necessary and sufficient conditions for the tensorial equations (10.7 - 9) to be satisfied are that the irreducible parts of the left and right-hand sides of each of the equations are equal (Toupin [462]).

According to this requirement the relations (10.6)

and (10.9) present no restrictions on the function ε, since the requirements are identically fulfilled, but the relations (10.6) and (10.7) present considerable restrictions.

On the left-hand side of (10.7) we have the symmetric part of the stress tensor, and hence the antisymmetric part of the right-hand side must vanish. This yields the set of three equations

$$\left[g^{i\ell}\left(\frac{\partial\varepsilon}{\partial x^{\ell}_{;L}}x^{j}_{;L} + \frac{\partial\varepsilon}{\partial x^{\ell}_{;KL}}x^{j}_{;KL} + \frac{\partial\varepsilon}{\partial d^{\ell}_{(\lambda);K}}d^{j}_{(\lambda);K}\right)\right]_{[ij]} = 0 . \tag{10.11}$$

If we compare this with (9.1.1), which followed from the principle of material frame indifference, we see that (10.11) is identical with (9.1.1). Accordingly, the internal energy must be a function of the form

$$\varepsilon = \varepsilon(C_{AB}, G_{CAB}, F_{\alpha AB}, \eta, X^{K}) . \tag{10.12}$$

To investigate the restrictions imposed by the symmetries of (10.8) we have first to find the irreducible parts of the tensor $m^{i(jk)} \equiv M^{ijk}$, knowing that $m^{ijk} = - m^{jik}$. According to the Appendix, (A2.26 - 29), the irreducible parts of the tensor M^{ijk} are

$$(10.13a) \qquad {}_{S}M^{ijk} = 0$$

$${}_{A}M^{ijk} = 0$$

$$
{}_{p}M^{ijk} = \frac{1}{6}(2m^{ijk} - m^{jki} - m^{kij}) ,
$$

(10.13b)

$$
{}_{\bar{p}}M^{ijk} = \frac{1}{6}(m^{ijk} + m^{jki} - 2m^{kij}) = -{}_{p}M^{kij} .
$$

Hence, the right-hand side of (10.8) has to satisfy 10 conditions $(10.13)_1$,

$$
\left(g^{il}\frac{\partial \varepsilon}{\partial x^{l}_{;KL}} x^{j}_{;K} x^{k}_{;L}\right)_{(ijk)} = 0 , \tag{10.14}
$$

and one condition $(10.13)_2$,

$$
\left(g^{il}\frac{\partial \varepsilon}{\partial x^{l}_{;KL}} x^{j}_{;K} x^{k}_{;L}\right)_{[ijk]} = 0 , \tag{10.15}
$$

and the tensor $m^{i(jk)}$ has only 8 independent components.

Owing to the symmetry of the gradients $x^{l}_{;KL} = x^{l}_{;LK}$ (10.15) is identically satisfied.

Relations (10.14) represent an additional system of 10 partial differential equations which must be satisfied simultaneously with the system (10.11). According to the definitions of the tensors $\underset{\sim}{C}$, $\underset{\sim}{G}$ and $\underset{\sim}{F}_{(\alpha)}$, (9.1.2-4), it is obvious that (10.14) will yield restrictions only on the tensor $\underset{\sim}{G}$. It may be directly verified that the system (10.14) is satisfied by the material tensor

(10.16) $$D_{ABC} \equiv G_{C[BA]} = C_{C[A,B]} \,.$$

Hence, the specific internal energy ε is an arbitrary function of the tensors $\underset{\sim}{C}, \underset{\sim}{D}, \underset{\sim}{F}_{(\alpha)}$ and of θ and X^K. For homogeneous materials ε does not depend on X^K,

(10.17) $$\varepsilon = \varepsilon(C_{AB}, D_{ABC}, F_{\alpha AB}, \eta) \,.$$

To write the mechanical constitutive equations (10.7-9) we have to perform the differentiations of the internal energy function considering it as an arbitrary function of the form (10.17), which gives for the derivatives the following expressions:

(10.18) $$\frac{\partial \varepsilon}{\partial x^{\ell}_{;L}} = \frac{\partial \varepsilon}{\partial C_{AB}} \frac{\partial C_{AB}}{\partial x^{\ell}_{;L}} + \frac{\partial \varepsilon}{\partial D_{ABC}} \frac{\partial D_{ABC}}{\partial x^{\ell}_{;L}} + \frac{\partial \varepsilon}{\partial F_{\alpha AB}} \frac{\partial F_{\alpha AB}}{\partial x^{\ell}_{;L}} \,,$$
$$\frac{\partial \varepsilon}{\partial x^{\ell}_{;KL}} = \frac{\partial \varepsilon}{\partial D_{ABC}} \frac{\partial D_{ABC}}{\partial x^{\ell}_{;KL}} \,,$$
$$\frac{\partial \varepsilon}{\partial d^{\ell}_{(\lambda);K}} = \frac{\partial \varepsilon}{\partial F_{\alpha AB}} \frac{\partial F_{\alpha AB}}{\partial d^{\ell}_{(\lambda);K}} \,.$$

According to (10.11), the equation for the symmetric part of the stress tensor becomes now

(10.19) $$t^{(ij)} = \varrho\left[g^{i\ell}\left(\frac{\partial \varepsilon}{\partial x^{\ell}_{;L}} x^{j}_{;L} + \frac{\partial \varepsilon}{\partial x^{\ell}_{;KL}} x^{j}_{;KL}\right)\right]_{(ij)} \,,$$

and the complete set of the mechanical constitutive relations is

$$t^{(ij)} = \varrho\left(2\frac{\partial\varepsilon}{\partial C_{KL}}x^{i}_{;K}x^{j}_{;L} + \frac{\partial\varepsilon}{\partial D_{KLM}}x^{(i}_{;K}x^{j)}_{;LM} + \frac{\partial\varepsilon}{\partial F_{\alpha KL}}x^{i}_{;K}d^{j}_{(\alpha);L}\right), \tag{10.20}$$

$$m^{i(jk)} = -\varrho\frac{\partial\varepsilon}{\partial D_{KLM}}x^{i}_{;K}x^{(j}_{;L}x^{k)}_{;M} , \tag{10.21}$$

$$h^{(\lambda)ij} = \varrho\frac{\partial\varepsilon}{\partial F_{\lambda KL}}x^{i}_{;K}x^{j}_{;L} . \tag{10.22}$$

For applications it is advantageous to substitute the deformation tensor $\underset{\sim}{C}$ by the strain tensor $\underset{\sim}{E}$ (3.10). It is also possible to represent the tensor $\underset{\sim}{D}$ in terms of the strain gradients,

$$D_{ABC} = 2E_{C[A,B]} . \tag{10.23}$$

From the constitutive relations (10.20–22) we see that the symmetric part of the stress tensor is affected by the strain of position, by the strain gradients and by the deformations of the directors, but couple-stresses depend (explicitly) only on the strain gradients, and the director stresses depend explicitly only on the deformations of the directors.

It is to be explicitly mentioned that in the thermodynamical approach to the constitutive relations the couple stress tensor remains indetermined. Out of its nine components only eight appear in the equation of energy balance and only eight are determined by the constitutive relations.

So far, except in the theory of dislocations (Kröner and Hehl [200], Stojanović [419, 421], Stojanović and Djurić [425]) the general relations (10.20-22) were not used in the applications. The applications are mostly concerned with more special classes of materials, i.e. with materials of grade two (the strain gradient theory), and with different kinds of oriented (directed) materials. For the materials of grade two the internal energy is assumed to be of the form

$$\varepsilon = \varepsilon(\underset{\sim}{C}, \underset{\sim}{D}, \eta), \tag{10.24}$$

and for the oriented materials of the form

$$\varepsilon = \varepsilon(\underset{\sim}{C}, \underset{\sim}{D}_{\alpha}, \eta). \tag{10.25}$$

In the section 4. we have already discussed the compatibility conditions for the deformation tensor $\underset{\sim}{C}$. To obtain the compatibility conditions for the tensor $\underset{\sim}{D}$ we shall use the commutativity of the covariant differentials in the Euclidean space. From (10.16) we obtain after differentiation

$$C_{CA,BD} = 2D_{ABC,D} + C_{CB,AD} .$$

Eliminating the derivatives of the tensor $\underset{\sim}{C}$ we obtain

$$D_{ABC,D} + D_{BDC,A} + D_{DAC,B} = 0 . \tag{10.26}$$

From the definition (9.1.4) of the tensors $\underset{\sim}{F}_{\alpha}$ we find

$$d_{(\alpha)i,j} = F_{\alpha AB} X^{A}_{;i} X^{B}_{;j} . \qquad (10.2.7)$$

Assuming that $X^{A}_{;i}$ are deformation gradients, we may write

$$d_{(\alpha)i,j} = F_{\alpha ij} ,$$

and for the spatial components of the tensors $\underset{\sim}{F}_{\alpha}$ we have

$$F_{\alpha i[j,k]} = 0 . \qquad (10.2.8)$$

Now, from (10.27) we have

$$d_{(\alpha)i,jk} = F_{\alpha AB,C} X^{A}_{;i} X^{B}_{;j} X^{C}_{;k} + F_{\alpha AB}(X^{A}_{;ik} X^{B}_{;j} + X^{A}_{;i} X^{B}_{;jk}) ,$$

and obviously

$$F_{\alpha [AB,C]} = 0 , \qquad (10.2.9)$$

which represents the compatibility conditions for the tensors $\underset{\sim}{F}_{\alpha}$.

10.1 A Principle of Virtual Work and Boundary Conditions

To derive the boundary conditions for elastic polar materials we shall generalize the principle of virtual work used by Toupin [462] for static equilibrium in the theory of elastic materials of grade two. In a slightly more general form this principle was also applied to generalized Cosserat con tinua by Stojanović and Djurić [426] .

We assume the principle of virtual work in the form

$$\delta T + \delta E = \delta w , \tag{10.1.1}$$

where δT is the virtual work of inertial forces, δE is the first variation of the internal energy and δw is the virtual work of all body and contact forces acting on a part v of a body. At the points of the boundary s of v the <u>normal derivatives</u> $D\delta x^i$ and $D\delta d^i_{(\alpha)}$ of (by assumption) independent variations δx^i and $\delta d^i_{(\alpha)}$ are to be considered also as independent.

In general, it may be assumed that the boundary s consists of a finite number of surfaces $\mathfrak{J}$ bounded by curves $\mathfrak{C}$. The boundary curves represent edges.

The gradients $\varphi_{,k}$ of a function φ, defined in the interior and on the boundary of v, may be decomposed on the boundary of v into the surface gradient $D_k\varphi$ and the normal gradient $D\varphi$,

$$\varphi_{,k} = D_k\varphi + n_k D\varphi , \tag{10.1.2}$$

where $\underset{\sim}{n}$ is the unit normal to the boundary surface s . Toupin introduced a three-dimensional extension of the second fundamental tensor $\underset{\sim}{b}$ of a surface by* (see foot-note next page)

$$b_{ij} = -D_i n_j = -D_j n_i . \tag{10.1.3}$$

For any smooth tensor field $f\dots$ defined at points of a smooth surface $\mathfrak{J}$ Toupin introduced the integral identity

$$\int_{\mathfrak{J}} D_i f\dots n_j \, ds = \int_{\mathfrak{J}} (b^k{}_k n_i n_j - b_{ij}) f\dots ds + \oint_{\mathfrak{C}} m_i n_j f\dots dl \,, \qquad (10.1.4)$$

where $\underset{\sim}{m} = \underset{\sim}{\tau} \times \underset{\sim}{n}$ and $\underset{\sim}{\tau}$ is the unit tangent to $\mathfrak{C}$, and dl is the scalar line element of $\mathfrak{C}$.

If the integral transformation (10.1.4) is applied to all surfaces $\mathfrak{J}$, i.e. to the whole boundary s of v, one gets

$$\int_{S} D_i f\dots n_j \, ds = \oint_{S} (b^k{}_k n_i n_j - b_{ij}) f\dots ds + \int_{c} [m_i n_j f\dots] \, dl \,, \qquad (10.1.5)$$

where $[\quad]$ represents the jumps of the enclosed quantity when an edge is approached from either side. We assume that the boundary s of v has no edge and that $f\dots$ is smooth throughout s, so that the line integral in (10.1.5) vanishes.

For the virtual work of inertia forces we assume the expression

$$\delta T = \int_{v} \varrho (\ddot{x}^i \delta x_i + i^{\lambda\mu} \ddot{d}^i_{(\lambda)} \delta d_{(\mu)i}) \, dv \,, \qquad (10.1.6)$$

---

* Let $u^\alpha, \alpha = 1,2$ be coordinates on S, and the equations of the surface are $x^i = x^i(u^\alpha)$. From (10.1.3) it follows that

$$b_{ij} x^i_{;\alpha} x^j_{;\beta} = n_j x^j_{;\alpha\beta} \equiv b_{\alpha\beta} \,,$$

where $b_{\alpha\beta}$ is the second fundamental tensor and $x^j_{;\alpha\beta}$ are covariant derivatives of $x^j_{;\alpha}$ with respect to the surface metric. It is to be noted that for the points on the surface $n_j x^j_{;\alpha} = 0$.

and for the variation of the internal energy we may write

$$\text{(10.1.7)} \qquad \delta E = \int_v \varrho\left(\frac{\partial \mathcal{E}}{\partial x^k_{;K}}\delta x^k_{;K} + \frac{\partial \mathcal{E}}{\partial x^k_{;KL}}\delta x^k_{;KL} + \frac{\partial \mathcal{E}}{\partial d^k_{(\lambda);K}}\delta d^k_{(\lambda);K}\right) dv \,.$$

Since the spatial coordinates only are subject to variations we shall use the following relations:

$$\delta x^k_{;K} = (\delta x^k)_{,m}\, x^m_{;K}$$

$$\text{(10.1.8)} \qquad \delta x^k_{;KL} = [(\delta x^k)_{,m}\, x^m_{;K}]_{;L} = (\delta x^k)_{;m\ell}\, x^m_{;K} x^\ell_{;L} + (\delta x^k)_{,m}\, x^m_{;KL}$$

$$\delta d^k_{(\lambda);K} = (\delta d^k_{(\lambda)})_{,m}\, x^m_{;K} \,,$$

and (10.1.7) may be rewritten in the form

$$\text{(10.1.9)} \qquad \partial E = \int_v \varrho\left[\left(\frac{\partial \mathcal{E}}{\partial x^k_{;K}} x^m_{;K} + \frac{\partial \mathcal{E}}{\partial x^k_{;KL}} x^m_{;KL}\right)(\delta x^k)_{,m} + \frac{\partial \mathcal{E}}{\partial x^k_{;KL}} x^m_{;K} x^\ell_{;L}(\delta x^k)_{,m\ell} + \frac{\partial \mathcal{E}}{\partial d^k_{(\lambda);K}} x^m_{;K}(\delta d^k_{(\lambda)})_{,m}\right] dv \,.$$

For the sake of brevity in writing let us introduce the notation

$$A_k^{\cdot m} = \varrho\left(\frac{\partial \mathcal{E}}{\partial x^k_{;K}} x^m_{;K} + \frac{\partial \mathcal{E}}{\partial x^k_{;KL}} x^m_{;KL}\right) ,$$

$$\text{(10.1.10)} \qquad B_k^{\cdot m\ell} = \varrho \frac{\partial \mathcal{E}}{\partial x^m_{;KL}} x^m_{;K} x^\ell_{;L} \,,$$

$$P^{(\lambda)\cdot m}_{\cdot k} = \varrho \frac{\partial \mathcal{E}}{\partial d^k_{(\lambda);K}} x^m_{;K} \,,$$

and

$$\mathfrak{J}_1 = \int_v A_k^{\cdot m}(\delta x^k)_{,m}\,dv \,,$$

$$\mathfrak{J}_2 = \int_v B_k^{\cdot m\ell}(\delta x^k)_{,m\ell}\,dv \,, \tag{10.1.11}$$

$$\mathfrak{J}_3 = \int_v P^{(\lambda)\cdot m}_{\cdot k}(\delta d^k_{(\lambda)})_{,m}\,dv \,.$$

For $\mathfrak{J}_1$ we have

$$\begin{aligned}\mathfrak{J}_1 &= \int_v [(A_k^{\cdot m}\delta x^k)_{,m} - A_{k,m}^{\cdot m}\delta x^k]dv \\ &= \oint_S A_k^{\cdot m}\delta x^k n_m ds - \int_v A_{k,m}^{\cdot m}\delta x^k dv \,.\end{aligned} \tag{10.1.12}$$

$\mathfrak{J}_2$ may be written in the form

$$\mathfrak{J}_2 = \int_v [(B_k^{\cdot m\ell}\delta x^k)_{,m\ell} + B_{k\ ,m\ell}^{\cdot m\ell}\delta x^k]dv - 2\oint_S B_{k\ \ ,m}^{\cdot(m\ell)}\delta x^k n_\ell ds \,. \tag{10.1.13}$$

Since we may write

$$\mathfrak{J}_2' = \int_v (B_k^{\cdot m\ell}\delta x^k)_{,m\ell}\,dv = \oint (B_k^{\cdot m\ell}\delta x^k)_{,m} n_\ell ds \,,$$

applying the integral identity (10.1.5) this becomes

$$\mathfrak{J}_2' = \oint_S \Big\{[DB_k^{\cdot m\ell} n_m n_n + (b^t_{\ t} n_m n_\ell - b_{m\ell})B_k^{\cdot m\ell}]\delta x^k +$$

$$+ B_k^{\cdot m\ell} n_m n_\ell (D\delta x^k)\Big\} ds \,,$$

and for $\mathfrak{J}_2$ we definitively have

$$\begin{aligned}\mathfrak{J}_2 &= \int_v B_{k\ \ ,m\ell}^{\cdot m\ell}\,\delta x^k dv \,+ \\ &+ \oint_S \Big\{\big[DB_k^{\cdot m\ell} n_m n_\ell + (b_t^t n_m n_\ell - b_{m\ell})B_k^{\cdot m\ell} - 2B_{k\ \ ,m}^{\cdot(m\ell)} n_\ell\big]\delta x^k + \\ &+ B_k^{\cdot m\ell} n_m n_\ell (D\delta x^k)\Big\} ds \; . \end{aligned} \tag{10.1.14}$$

For $\mathfrak{J}_3$ we obtain similarly

$$\mathfrak{J}_3 = \oint_S P^{(\alpha)\cdot m}_{\cdot k}\,\delta d^k_{(\alpha)} n_m\, ds - \int_v P^{(\alpha)\cdot m}_{\cdot k\ ,m}\,\delta d^k_{(\alpha)} dv \; . \tag{10.1.15}$$

Collecting the results we obtain for δE the expression

$$\begin{aligned}\delta E &= \int_v \big[(-A_{k\ ,m}^{\cdot m} + B_{k\ \ ,m\ell}^{\cdot m\ell})\delta x^k - P^{(\alpha)\cdot m}_{\cdot k,m}\,\delta d^k_{(\alpha)}\big] dv \,+ \\ &+ \oint_S \Big\{\big[A_k^{\cdot m} n_m + (DB_k^{\cdot m\ell}) n_m n_\ell + (b_t^t n_m n_\ell - b_{m\ell})B_k^{\cdot m\ell} - 2B_{k\ \ ,m}^{\cdot m\ell} n_\ell\big]\delta x^k + \\ &+ P^{(\alpha)\cdot m}_{\cdot k} n_m \delta d^k_{(\alpha)} + B_k^{\cdot m\ell} n_m n_\ell (D\delta x^k)\Big\} ds \; . \end{aligned} \tag{10.1.16}$$

According to the form of (10.1.16) it is natural to assume for the virtual work δw the expression

$$\begin{aligned}\delta w &= \int_v (L_k \delta x^k + S_k^{(\alpha)} \delta d^k_{(\alpha)}) dv \,+ \\ &+ \oint_S \big[M_k \delta x^k + N_k (D\delta x^k) + T_k^{(\alpha)} \delta d^k_{(\alpha)}\big] ds \; . \end{aligned} \tag{10.1.17}$$

where $\underset{\sim}{L}$, $\underset{\sim}{M}$, $\underset{\sim}{N}$, $\underset{\sim}{S}^{(\alpha)}$ and $\underset{\sim}{T}^{(\alpha)}$ are some generalized forces.

Introducing now $\delta T, \delta E$ and δw from (10.1.6, 16, 17) into (10.1.1) and assuming that the variations δX^k, $D\delta x^k$ and $\delta d^k_{(\alpha)}$ in v and on s are independent, we obtain the following relations: in $\check{v}$:

$$\varrho \ddot{x}^\ell - A^{\ell m}{}_{,m} + B^{\ell mn}{}_{,mn} = L^\ell , \tag{10.1.18}$$

$$\varrho i^{\alpha\mu} \ddot{d}^\ell_{(\alpha)} - P^{(\alpha)\ell m}{}_{,m} = s^{(\alpha)\ell} , \tag{10.1.19}$$

on s :

$$A^{\ell m} n_m + (DB^{\ell mn}) n_m n_n + (b^t_t n_m n_n - b_{mn}) B^{\ell mn} - 2B^{\ell mn}{}_{,m} n_n = M^\ell , \tag{10.1.20}$$

$$P^{(\alpha)\ell m} n_m = T^{(\alpha)\ell} , \tag{10.1.21}$$

$$B^{\ell mn} n_m n_n = N^\ell . \tag{10.1.22}$$

From (10.8), (10.9), (10.19) and (10.1.10) we see that

$$A^{(ij)} = t^{(ij)} ,$$

$$B^{\ell mn} = -m^{\ell(mn)} , \tag{10.1.23}$$

$$P^{(\alpha)\ell m} = h^{(\alpha)\ell m} .$$

According to (10.11) we also have *

$$A^{[\ell m]} - d^{[\ell}_{(\alpha),p} h^{(\alpha)m]p} = 0 \,, \tag{10.1.24}$$

which substituted in (10.1.18) yields

$$\varrho \ddot{x}^{\ell} = t^{(\ell m)}{}_{,m} + m^{\ell(mn)}{}_{,mn} + d^{[\ell}_{(\alpha),p} h^{(\alpha)m]p} + L^{\ell} \,.$$

This, together with (10.1.19),

$$\varrho i^{\alpha\mu} \ddot{d}^{\ell}_{(\alpha)} = h^{(\mu)\ell m}{}_{,m} + s^{(m)\ell} \,,$$

represents the equations of motion. Here we may identify L^{ℓ} with $\varrho(f^{\ell} + \ell^{\ell m}{}_{,m})$, and $s^{\alpha(\ell)}$ with $\varrho k^{(\alpha)\ell}$. The boundary conditions follow from (10.1.20–22),

$$t^{(\ell m)} n_m + d^{[\ell}_{(\alpha),p} h^{(\alpha)m]p} n_m + Dm^{\ell(mn)} n_m n_n - (b^t_t n_m n_n - b_{mn}) m^{\ell(mn)} + 2m^{\ell(mn)}{}_{,m} n_n = M^{\ell} \,,$$

$$h^{(\alpha)\ell m} n_m = T^{(\alpha)\ell} \,, \tag{10.1.25}$$

$$-m^{\ell(mn)} n_m n_n = N^{\ell} \,.$$

---

* The equation (10.1.24) follows also from the requirement that δE is invariant under virtual rigid displacements. Let x^i be Cartesian coordinates. The virtual rigid displacements are $\delta x^k = a^k + \varepsilon^{kij} K_i x_j$ and $\delta d^k_{(\alpha)} = \varepsilon^{kij} K_i d_{(\alpha)j}$, where a^k and K_i are arbitrary constants. Introducing this into (10.1.14) and requiring that the energy of every part of the body is separately invariant under all rigid variations we obtain (10.1.24).

The generalized forces $\underset{\sim}{M}$, $\underset{\sim}{T}^{(\alpha)}$ and $\underset{\sim}{N}$ are certain surface tractions which are to be prescribed on the boundary of the body.

10.2 Elastic Materials of Grade Two

When the internal energy is a function of deformation gradients $x^k_{;K}$ and $x^k_{;KL}$ and of X^K and η only, the mechanical constitutive relations (10.20, 21) obtain the form

$$t^{(ij)} = \varrho\left(2\frac{\partial\varepsilon}{\partial C_{KL}}x^i_{;K}x^j_{;L} + \frac{\partial\varepsilon}{\partial D_{KLM}}x^{(i}_{;K}x^{j)}_{;LM}\right), \tag{10.2.1}$$

$$m^{i(jk)} = -\varrho\frac{\partial\varepsilon}{\partial D_{KLM}}x^i_{;K}x^{(j}_{;L}x^{k)}_{;M}. \tag{10.2.2}$$

According to the Appendix (A1.32), the couple-stress tensor m^{ijk} may be represented by the second order tensor $m_l^{\cdot k}$, and this tensor may be decomposed into its deviatoric and spherical part, where the deviatoric part is

$$\mu_l^{\cdot k} \equiv m_l^{\cdot k} - m_p^{\cdot p}\delta_l^k = \frac{1}{2}\varepsilon_{lij}m^{ijk} - \frac{1}{2}\varepsilon_{pqr}m^{pqr}\delta_l^k \tag{10.2.3}$$

or

$$m^{ijk} = \mu^{ijk} + m_p^{\cdot p}\varepsilon^{ijk}, \tag{10.2.4}$$

where

$$\mu^{ijk} = \varepsilon^{ijl}\mu_l{}^k . \tag{10.2.5}$$

In the constitutive relations (10.2.2) only the symmetric part $m^{i(jk)}$ of the couple-stress tensor appears, and from (10.2.4) we see that

$$m^{i(jk)} = \mu^{i(jk)} . \tag{10.2.6}$$

Since there are only eight independent components of the tensor $m^{i(jk)}$ (cf. 10.13), and since the deviator has only eight components (cf. App. (A2.4)), we may represent the deviator μ^{ijk} in terms of the tensor $m^{i(jk)}$,

$$\mu^{ijk} = \frac{2}{3}(2m^{i(jk)} + m^{k(ij)}) . \tag{10.2.7}$$

The invariant $\varepsilon_{ijk}m^{ijk} = m^k_{.k}$ of the couple-stress tensor remains undetermined since there are only eight constitutive equations (10.2.2), and also in the boundary conditions (10.1.25) only the symmetric part of the couple-stress tensor appears. According to Koiter [241] , without any loss in generality we may assume that $m_k^{.k}$ is equal to zero.

The tensor D_{KLM} is antisymmetric in K and L and if we introduce the second-order material tensor

$$D^N_{.M} = \frac{1}{2}\varepsilon^{NKL}D_{KLM} , \tag{10.2.8}$$

the constitutive equations (10.2.1) obtain the form

$$t^{(ij)} = \varrho\left(\frac{\partial\varepsilon}{\partial E_{KL}} x^{i}_{;K} x^{j}_{;L} + \frac{1}{2}\frac{\partial\varepsilon}{\partial D^{N}_{\cdot M}} \varepsilon^{NKL} x^{(i}_{;K} x^{j)}_{;LM}\right), \qquad (10.2.9)$$

where we have used (3.10), and for the deviator $\mu_{l}^{\cdot k}$ we get from (10.2.3, 7, 8,) the relation

$$\mu_{l}^{\cdot k} = -\frac{1}{3}\varrho \frac{\partial\varepsilon}{\partial D^{N}_{\cdot M}} \varepsilon^{NKL} \varepsilon_{ijl} x^{i}_{;K} x^{(j}_{;L} x^{k)}_{;M} . \qquad (10.2.10)$$

For isotropic materials the internal energy must be a function of isotropic invariants (see App. section A2) of the tensors $\underset{\sim}{E}$ and $\underset{\sim}{D}$,

$$\varepsilon = \varepsilon(I_E, II_E, III_E, {}^{1}II_D, {}^{2}II_D, {}^{2}III_{ED}, {}^{3}III_{ED}, {}^{4}III_{ED}, \ldots) . \qquad (10.2.11)$$

Teodosiu [449 - 453] applied the general theory of elastic materials of grade two to media with internal and initial stresses and particularly to the determination of internal stresses produced by dislocations. He also considered a more general theory in which the couple-stress tensor is not undetermined. A proposal for such a generalization was already given by Toupin [462] on the basis of the analysis of the boundary conditions $(10.1.25)_3$. From the antisymmetry of the couple-stress tensor it follows that the traction $\underset{\sim}{N}$ has to be orthogonal to the boundary surface, $\underset{\sim}{N}\cdot\underset{\sim}{n} = 0$, but this requirement for the traction $\underset{\sim}{N}$ is without a physical motivation. For that reason Toupin proposed a more general theory in which the complete couple-stress tensor would be determined.

For infinitesimal deformations we may assume that the coordinates X^K and x^k coincide in the reference configuration, such that

$$x^k = X^K \delta^k_K + u^k ,$$

$$x^k_{;L} = \delta^k_L + u^k_{,\ell} \delta^\ell_L , \tag{10.2.12}$$

$$x^k_{;LM} = u^k_{,\ell m} \delta^\ell_L \delta^m_M ,$$

where $\underset{\sim}{u}$ is an infinitesimal displacement. The deformation tensors in the linear approximation are

$$E_{KL} \approx e_{k\ell} \delta^k_K \delta^\ell_L = u_{(k,\ell)} \delta^k_K \delta^\ell_L ,$$

(10.2.13)

$$D_{KLM} \approx D_{k\ell m} \delta^k_K \delta^\ell_L \delta^m_M ,$$

where

$$D_{k\ell m} = 2 e_{m[k,\ell]} = 2 w_{k\ell,m} , \tag{10.2.14}$$

$$w_{k\ell} = u_{[k,\ell]} . \tag{10.2.15}$$

It is accustomed, however, to represent the third-order tensors $\underset{\sim}{\mu}$ and $\underset{\sim}{D}$ by their second-order duals. Since the rotation tensor $w_{k\ell}$ may be represented by the vector $w^i = \frac{1}{2} \varepsilon^{ik\ell} w_{k\ell}$, we may put

$$k_{ij} = w_{i,j} , \tag{10.2.16}$$

and the linear constitutive relations may be written in the form

$$
\begin{aligned}
t^{(ij)} &= C_1^{ijkl} e_{kl} + C_2^{ijkl} k_{kl} \,, \\
\mu^{ij} &= M_1^{ijkl} e_{kl} + M_2^{ijkl} k_{kl} \,.
\end{aligned} \tag{10.2.17}
$$

For isotropic materials the fourth-order tensors $\underset{\sim}{C}$ and $\underset{\sim}{M}$ are linear combinations of the fundamental tensors such that

$$
\begin{aligned}
C_\nu^{ijkl} &= \alpha_\nu g^{ij} g^{kl} + \beta_\nu g^{ik} g^{jl} + \gamma_\nu g^{il} g^{jk} \,, \\
M_\nu^{ijkl} &= a_\nu g^{ij} g^{kl} + b_\nu g^{ik} g^{jl} + c_\nu g^{il} g^{jk} \quad (\nu = 1,2) \,,
\end{aligned} \tag{10.2.18}
$$

Since the constitutive relations (10.2.17) for isotropic materials have to be invariant under the full orthogonal group of transformations, we shall obtain them substituting the elasticity tensors from (10.2.18) into (10.2.17).

In the linear theory we may assume that the density ϱ is approximatively equal to the density in the reference configuration, $\varrho \approx \varrho_0$.

For isotropic materials in this approximation the internal energy function may be approximated by a quadratic polynomial in the isotropic invariants I_e, II_e and ${}^1II_D$, ${}^2II_D$ of the tensors $\underset{\sim}{e}$ and $\underset{\sim}{D}$, and it may be written in the form (Koiter [241])

(10.2.19) $$\varrho_0\mathcal{E} = G\left[\frac{\nu}{1-2\nu}I_e^2 + e^i_{\cdot j}e^j_{\cdot i} + 2\ell^2(k^{\cdot i}_{\cdot j}k^{\cdot j}_{i} + \eta k^{\cdot i}_{\cdot j}k^{\cdot j}_{\cdot i})\right] ,$$

where G is the shear modulus, ν is the Poisson ratio and $2G\ell^2$ and $2\eta G\ell^2$ are two additional new elastic constants. The constant ℓ has the dimension of length and is called the characteristic length of the material. η is a non-dimensional number.

The constitutive relations (10.2.9, 10) may be written now in the form

(10.2.20) $$\begin{aligned} t^{(ij)} &= \varrho_0\frac{\partial\mathcal{E}}{\partial e_{ij}} = 2G\left(e^{ij} + \frac{\nu}{1-2\nu}I_e g^{ij}\right) , \\ \mu^{ij} &= \varrho_0\frac{\partial\mathcal{E}}{\partial k_{ji}} = 4G\ell^2(k^{ij} + \eta k^{ji}) . \end{aligned}$$

These relations were obtained by Aero and Kuvshinskii [5] in 1960. Grioli [180] studied the non-linear theory and in the linearization he obtained the similar expressions, but he neglected the terms involving η. Mindlin and Tiersten [283] considered the linear constitutive equations as a result of linearization of the relations derived by Toupin, and they applied the linear theory to a number of problems in vibrations and stress concentration (cf. also Mindlin [287]). One of the most interesting effects of couple-stresses is its influence on the stress concentration factor which appears to be a function of the characteristic length ℓ and to be less than what is usually assumed in the non-polar theories to be its value. For detailed study of the influence of couple-stresses in linear elasticity we refer the reader, among others, to the papers by Mindlin and Tiersten [283], Mindlin [284], [287], Mindlin and Eshel [288], Koiter [241], Neuber [324], and, for the problems of stress concentration, to the book by Savin [387] which appeared in 1968 and where detailed references may be found.

Lomakin [275] applied Lagrange's variational principle to derive various boundary conditions. He also completed the theory proving the validity of the principle of minimum potential energy, generalizing Clapeyron's theorem for the strain energy and proving the uniqueness theorem.

Within the theory of materials of grade two (or, within the strain-gradient theory) a generalization of Rivlin's method for the construction of general solutions in non-linear elasticity was presented by Stojanović and Blagojević [424] and by Blagojević [33, 34]. It is found that owing to the influence of couple-stresses the Poynting effect, which is in the non-linear theory of elasticity attributed to the second-order terms, appears as an effect of the first order in hemitropic materials.

A very fine and general synthesis of work of Grioli, Aero and Kuvshinskii, Bressan [47] and other authors is presented by Galletto [107]

10.3 The Elastic Cosserat Continuum

When the influence of the strain gradients in the internal energy function is neglected, according to (10.20-22) the couple-stress tensor $\underset{\sim}{m}$ will vanish and the constitutive relations obtain the form

$$t^{(ij)} = \varrho\left(\frac{\partial \varepsilon}{\partial E_{KL}} x^{i}_{;K} x^{j}_{;L} + \frac{\partial \varepsilon}{\partial F_{\alpha KL}} x^{i}_{;K} d^{j}_{(\alpha);L}\right), \tag{10.3.1}$$

$$h^{(\lambda)ij} = \varrho \frac{\partial \varepsilon}{\partial F_{(\lambda)KL}} x^{i}_{;K} x^{j}_{;L}. \tag{10.3.2}$$

The directors in a Cosserat medium represent rigid triads and therefore we may assume that in the initial (reference) configuration the directors $D^{K}_{(\alpha)}$ coincide with the base vectors of a Cartesian system of reference X^{K}, i.e.

$$\underset{\sim}{D}_{(\alpha)} = D^{K}_{(\alpha)} \underset{\sim}{e}_{K}, \quad D^{K}_{(\alpha)} = \delta^{K}_{\alpha}. \tag{10.3.3}$$

For infinitesimal deformation we may write

$$\begin{aligned} x^{k} &= X^{K} \delta^{k}_{K} + u^{k}, \\ \underset{\sim}{d}_{(\alpha)} &= \underset{\sim}{D}_{(\alpha)} + \underset{\sim}{\Omega} \times \underset{\sim}{D}_{(\alpha)} \end{aligned} \tag{10.3.4}$$

or

$$d_{(\alpha)}{}^{k} = \delta_{\alpha}{}^{k} + \Omega_{\alpha}{}^{\cdot k} \tag{10.3.5}$$

where $\underset{\sim}{u}$ is an infinitesimal displacement vector, and $\underset{\sim}{\Omega}$ is an independent rotation of the director triads. However,

$$x^k_{;K} = \delta^k_K + u^k_{,\ell}\delta^\ell_K ,$$
$$d^k_{(\alpha);\ell} = \Omega^{\cdot k}_{\alpha,\ell}\delta^\ell_L , \tag{10.3.6}$$

and the deformation tensors are

$$E_{k\ell} \approx u_{(k,\ell)}\delta^k_K\delta^\ell_L$$
$$F_{\alpha KL} \approx \Omega_{\alpha k,\ell}\delta^k_K\delta^\ell_L \equiv x_{\alpha k\ell}\delta^k_K\delta^\ell_L . \tag{10.3.7}$$

thus, we may consider as the constitutive variables the strain tensor $e_{k\ell}$ and the gradients of rotation $x_{\alpha k\ell} = -x_{k\alpha\ell}$ or

$$x^m_{\cdot\ell} \equiv \frac{1}{2}e^{\alpha km}\,\Omega_{\alpha k,\ell} . \tag{10.3.8}$$

From (10.28) we easily obtain the compatibility conditions for the tensor $x_{\alpha k\ell}$. From (10.3.7) we have

$$F_{\alpha[KL,M]} \approx x_{\alpha[k\ell,m]}\delta^k_K\delta^\ell_L\delta^m_M .$$

Since the indices k,ℓ,m here must have different values, there are just three independent relations, which may be written in the form

$$e^{k\ell m}x_{\alpha k\ell,m} = 0 .$$

Using now the antisymmetry of $x_{\alpha k l} = -x_{k \alpha l}$, and writing

$$x_{\alpha k l} = e_{\alpha k t} x^{t}_{\cdot l}$$

we find

$$x^{l}_{\cdot [l,m]} = 0 \; .$$

This is, however, identically satisfied, since from $(10.3.7)_2$ we see that the relation

$$x^{m}_{\cdot [l,n]} = 0$$

represents the compatibility condition. In this context we also refer the reader to the compatibility conditions for micromorphic elastic media derived by Eringen [134] .

The constitutive relations (10.3.1, 2) for $\varrho \approx \varrho_0$ become now

$$\text{(10.3.9)} \qquad \begin{aligned} t^{(ij)} &= \varrho_0 \left(\frac{\partial \varepsilon}{\partial e_{ij}} + \frac{\partial \varepsilon}{\partial x_{\lambda i l}} x^{\; j}_{\lambda \cdot l} \right), \\ h^{(\lambda) ij} &= \varrho_0 \frac{\partial \varepsilon}{\partial x_{\lambda ij}} \; . \end{aligned}$$

however, $x_{\lambda ij}$ is an antisymmetric tensor and the index λ is of the tensorial character. Applying (10.3.8) we may now write

$$\text{(10.3.10a)} \quad t^{(ij)} = \varrho_0 \left(\frac{\partial \varepsilon}{\partial e_{ij}} + \frac{\partial \varepsilon}{\partial x^{m}_{\cdot n}} x^{m}_{\cdot n} \delta^{ij} - \frac{\partial \varepsilon}{\partial x^{m}_{\cdot n}} x^{i}_{\cdot n} \delta^{jm} \right)$$

$$h_{\lambda}^{\cdot j} = \varrho_0 \frac{\partial \varepsilon}{\partial x^{i}_{;j}} , \tag{10.3.10b}$$

where

$$h_{\nu}^{\cdot j} = \frac{1}{2} \varepsilon_{i\lambda n} h^{\lambda n j} . \tag{10.3.11}$$

The internal energy ε may be approximated now by a quadratic polynomial,

$$\varrho_0 \varepsilon = G\left[\frac{\nu}{1-2\nu} I_e^2 + e^{i}_{j} e^{j}_{i} + 2\overset{*}{l}{}^2 (x^{i}_{;j} x^{;j}_{i} + \eta^* x^{i}_{;j} x^{j}_{;i})\right] \tag{10.3.12}$$

and the linear consitutive relations have the form completely analogous to (10.2.20),

$$\begin{aligned} t^{(ij)} &= 2G\left(e^{ij} + \frac{\nu}{1-2\nu} I_e g^{ij}\right) , \\ h_{j}^{\cdot i} &= 4G\overset{*}{l}{}^2 (x_{j}^{;i} + \overset{*}{\eta} x^{i}_{;j}) . \end{aligned} \tag{10.3.13}$$

Here again we have a "characteristic length" $\overset{*}{l}$ of the material, and a nondimensional constant $\overset{*}{\eta}$.

The linear theory of elasticity of Cosserat materials is studied extensively by Schäfer [390–395] , who also elaborated a method for solving the equilibrium problems in terms of the stress-functions [394] , and applied the theory to the theory of dislocations * [396–398].

---

* I mostly appreciate the late Prof. Schäfer's kindness to put at my disposal his yet unpublished results on the dislocation theory in the Cosserat continuum.

The theory of non-symmetric elasticity developed since 1960 by Aero, Bul'gin and Kuvshinskii [4-6, 52,260] is based on the assumption that particles of a medium may suffer rotations independent of the displacements, which makes their theory to be, in fact, a theory of Cosserat media.

The equations of motion $(7.1.6)_{2,4}$ in the linearized theory of Cosserat continua obtain the form

$$\varrho \ddot{x}^i = t^{ij}_{\ ,j} + \varrho f^i$$

(10.3.14)

$$\varrho I^{\ell[i} \dot{\omega}_\ell^{\ j]} = t^{[ji]} + H^{[ij]k}_{\ \ \ ,k} + \varrho d^{[i}_{(\alpha)} k^{(\alpha)j]} ,$$

where the hyperstress tensor H^{ijk} defined by $(7.25)_4$, appears only as an antisymmetric tensor.

The moments of director forces appear here in the form of body couples. The effect of hyperstresses in the linear theory of an elastic Cosserat continuum is obviously the same as the effect of couple-stresses in the strain-gradient theory. For that reason many authors consider both kinds of "materials" as Cosserat materials, or simply as materials with couple-stresses without making any distinction between the two kinds of materials.

Transvecting the equation (10.3.14) with ε_{mij} and representing the rotation tensor ω^{nj} by the rotation vector $\omega_t = \frac{1}{2}\varepsilon_{tnj}\omega^{nj}$, we obtain

$$\varrho j^{\ t}_{m}\dot{\omega}_t = \varepsilon_{mij} t^{ji} + H_{m,k}^{\ \ \cdot k} + \varrho M_m , \tag{10.3.15}$$

where (see Section 7.1)

$$j^{t}_{m} = I^{n}_{n}\delta^{t}_{m} - I^{t}_{m}, \quad H^{\cdot k}_{m} = \varepsilon_{mij}H^{ijk} ,$$
$$M_{m} = \varepsilon_{mij}d^{i}_{(\alpha)}k^{(\alpha)j} . \qquad (10.3.16)$$

For microisotropic media by (7.1.10) we have

$$j^{t}_{m} = j\delta^{t}_{m} \qquad (10.3.17)$$

and the equation (10.3.15) obtains the form often used by various authors in the linearized theories.

10.4 Elastic Materials with Microstructure

a) Micromorphic and micropolar materials.- The basic theory is developed by Eringen and Suhubi [123-130, 137-139, 442] . It is assumed that for the microelements are valid the Cauchy laws of motion,

$$\varrho' a'^{i} = t'^{ij}_{\ ,j} + \varrho f'^{i} ,$$
$$t'^{ij} = t'^{ji} \qquad (10.4.1)$$

where primes denote that the quantities are related to microelements. For macromaterial the corresponding quantities are obtained through the averaging, e.g.

(10.4.2) $$\int_{ds} t'^{ij} ds'_j = t^{ij} ds_j , \quad \int_{dv} \varrho' f'^i dv' = \varrho f^i dv ; \quad \text{etc.}$$

The stress and volume moments are defined by the relations

(10.4.3) $$\begin{aligned} \int_{ds} t'^{ij} \xi'^k ds'_j &= \lambda^{ijk} ds_j . \\ \int_{dv} \varrho' f'^i \xi'^j dv' &= \varrho l^{ij} dv , \end{aligned}$$

and λ^{ijk} represents the "first stress moment", which is not the same as the couple-stress. Further, in the relation

(10.4.4) $$\int_{dv} \varrho' a'^i \xi'^j dv' = \varrho \dot{\sigma}^{ij} dv$$

the quantity $\dot{\sigma}^{ij}$ is defined as the "inertial spin", and the symmetric tensor s^{ij}, defined by

(10.4.5) $$\int_{dv} t'^{ij} dv' = s^{ij} dv$$

represents the "microstress average".

The constitutive relations, according to our notation (cf. section 5.2) read

(10.4.6a) $$\begin{aligned} t^k_{.l} &= \varrho \frac{\partial \varepsilon}{\partial x^l_{;K}} x^k_{;K} , \\ s^k_{.l} &= \varrho \left(\frac{\partial \varepsilon}{\partial x^l_{;K}} x^k_{;K} + \frac{\partial \varepsilon}{\partial d^l_{(\lambda)}} d^k_{(\lambda)} + \frac{\partial \varepsilon}{\partial d^l_{(\lambda);K}} d^k_{(\lambda);K} \right) , \end{aligned}$$

$$\lambda^{k\cdot m}_{\cdot l} = \varrho \frac{\partial \varepsilon}{\partial d^{l}_{(\lambda);L}} x^{k}_{;L} d^{m}_{(\lambda)} . \qquad (\lambda = 1,2,3) \qquad (10.4.6b)$$

where it is assumed that the internal energy ε is a function of the mechanical constitutive variables

$$x^{k}_{;K} , \quad d^{l}_{(\lambda)} , \quad d^{l}_{(\lambda);K} . \qquad (10.4.7)$$

The stress moment $\underset{\sim}{\lambda}$ coincides with our hyperstress, and this theory may be regarded also as a theory of generalized Cosserat continua.

The difference between the general theory outlined in the section 10.3 and the theory of micromorphic continua is in the assumption that the internal energy depends explicitly on the components of the directors, and, also, in the assumed existence of two independent stresses - the macro-stress $\underset{\sim}{t}$ and the micro-stress average $\underset{\sim}{s}$.

In micropolar bodies the micro-elements are rigid. The directors in this case represent rigid triads and the theory reduces to the theory of elastic Cosserat media (10.3).

If we assume the internal energy ε to be a function of the variables (10.4.7) and of the specific entropy η ,

$$\varepsilon = \varepsilon(x^{l}_{;k} , d^{l}_{(\lambda)} , d^{l}_{(\lambda);K} , \eta) ,$$

and if we apply the principle of material frame indifference to obtain the equations which correspond to (9.17), we shall obtain that $\mathcal{E}$ is an arbitrary function of η and of the materials tensors

$$C_{KL} = g_{k\ell} x^k_{;K} x^\ell_{;L} ,$$

(10.4.8)
$$\Psi_{K\lambda} = g_{k\ell} x^k_{;K} d^\ell_{(\lambda)} ,$$

$$\Gamma_{K\lambda L} = g_{k\ell} x^k_{;K} d^\ell_{(\lambda);L} .$$

Owing to the symmetry of the tensor $\underset{\sim}{C}$ there are 42 independent integrals $\underset{\sim}{C}, \underset{\sim}{\Psi}$ and $\underset{\sim}{\Gamma}$ of the (three in number) equations (9.17). In the theory of micromorphic bodies there are only three directors and the greek indices are regarded also as material tensorial indices.

The tensors $\underset{\sim}{C}$ and $\underset{\sim}{\Gamma}$ are included in the general theory of the section 10, in which the tensor $\underset{\sim}{D}$ is to be omitted since $\mathcal{E}$ does not depend now on the second-order deformation gradients $x^k_{;KL}$. The tensor $\underset{\sim}{\Gamma}$ corresponds to $\underset{\sim}{F}_\lambda$. $\underset{\sim}{\Psi}$ and $\underset{\sim}{\Gamma}$ are called <u>micro-deformation tensors.</u>

As in the case of Cosserat materials in the section 10.3, for micropolar materials we may write

(10.4.9)
$$d^\ell_{(\lambda)} \approx \delta^\ell_\lambda + \omega_\lambda^{\cdot\ell}$$

and for the tensor $\underset{\sim}{\Psi}$ we have

(10.4.10)
$$\Psi_{K\lambda} \approx g_{\lambda K} + u_{\lambda;K} + \omega_{\lambda K} .$$

The corresponding strain tensor

$$\varepsilon_{k\ell} = (\Psi_{K\lambda} - g_{\lambda K})\delta^K_k\delta^\lambda_\ell = u_{k,\ell} + \omega_{k\ell} , \qquad (10.4.11)$$

is not symmetric. Its symmetric part coincides with the strain tensor corresponding to $\underset{\sim}{C}$, so that in the linear theory of micropolar bodies the state of strain is described in terms of the strain components

$$\varepsilon_{(k\ell)} = e_{k\ell} = u_{(k,\ell)} ,$$

$$\varepsilon_{[k\ell]} = w_{k\ell} + \omega_{k\ell} , \qquad (10.4.12)$$

$$\gamma_{k\ell m} = \omega_{k\ell,m} .$$

These measures of strain appear in the theories of Aero and Kuvshinskii [5, 6] and in many other linear theories of Cosserat media.

b) Microstructure.- The linear theory of elastic bodies with microstructure was developed by Mindlin [284, 285]. The continuum is composed of unit cells which have some properties of crystal lattices. The theory represents, in the mechanical sense, the linearized version of the theory of the generalized Cosserat continua with deformable directors (section 5.2). The directors represent microdeformations, and since there are only three directors in this theory we may put $d_{(\alpha)j} = \Psi_{\alpha j}$, where $\Psi_{\alpha j}$

are displacement-gradients in the micro-medium,

$$\psi_{ij} = \partial u'^{j}/\partial x'^{i} .$$

Denoting by x^i and u^i Cartesian coordinates and components of the macro-displacements, resp., the <u>relative deformation</u> is given by

$$\gamma_{ij} = \frac{\partial u_j}{\partial x^i} - \psi_{ij} ,$$

and the <u>macro-strain</u> by

$$\varepsilon_{ij} = \partial_{(i} u_{j)} .$$

Macro-deformation gradients are determined by the tensor

$$x_{ijk} = \partial_i \psi_{jk}$$

which represents the tensor of director-gradients.

The state of stress is described by the ordinary (Cauchy) stress t^{ij}, by the <u>relative stress</u> σ^{ij}, and by the <u>double stress</u> μ^{ijk}, such that (for $\varrho = \varrho_0 = 1$)

$$(10.4.13) \qquad t^{ij} = \frac{\partial \mathcal{E}}{\partial \varepsilon_{ij}} , \quad \sigma^{ij} = \frac{\partial \mathcal{E}}{\partial \gamma_{ij}} , \quad \mu_{ijk} = \frac{\partial \mathcal{E}}{\partial x_{ijk}} ,$$

and the equations of motion are

$$(10.4.14) \qquad \begin{aligned} (t^{ij} + \sigma^{ij})_{,j} + \varrho f^i &= \varrho \ddot{u}^i \\ \mu^{ijk}{}_{,i} + \sigma^{jk} + \phi^{jk} &= i^{jl} \ddot{\psi}_l{}^{k} . \end{aligned}$$

Φ^{jk} are certain <u>double forces</u>, and $\iota^{jl} = \frac{1}{3}\varrho(d^{jl})^2$ are certain inertial coefficients. The quantities d^{jl} depend on the "unit cell" of the medium considered. The symmetric part $\Psi_{(ij)}$ of the microdeformation represents the micro-strain, and the antisymmetric part is the micro-rotation, $\Psi_{[ij]} = \omega_{ij}$ (cf. Section 5.2).

This theory contains the linearized equations of Cosserat continua as a special case, and the linear version of the strain-gradient theory as a special case, too. Eringen [130] showed, however, that this theory coincides with the theory of micromorphic materials. The theory of Mindlin, however, is elaborated only in the linear version and it is difficult to say from the coincidence of two theories in their linear form if they agree in general, or they represent two different theories.

10.5 Incompatible Deformations

Under certain circumstances a field of stresses cannot be associated to a field of deformations which satisfies the compatibility conditions (see App. sections A4 and section 4). Such situations appear in thermoelasticity and in the theory of dislocations. In the classical linear thermoelasticity, in the Duhamel – Neumann law, it is assumed that the total strain $\underset{\sim}{e}$ which satisfies the compatibility conditions, is composed of two strains which do not satisfy these conditions, of an elastic strain $\underset{\sim}{e}^E$ which produces thermal stresses, and of a strain $\underset{\sim}{e}^T$ which depends on the distribution of temperature in a body. This idea was used in the linear theory of dislocations for the determination of internal stresses produced by dislocations (cf. Kröner [246]) and later it was generalized first in the theory of dislocations by Kröner and Seeger [247] . Günther [189] established a very important and interesting relation between the incompatibilities of the Cosserat continuum and the structural curvature of a dislocated crystal.

Stojanović, Djurić and Vujoshević [419-421, 429, 432-434, 475-478] developed a general theory of elastic incompatible deformations which was applied to thermoelasticity and dislocations [419, 421] .

The theory is based on the assumptions (see section 4.1) that the deformation gradients corresponding to a deformation $x^k = x^k(\underset{\sim}{X})$ of a body from an initial (and unstressed) configuration K_0 into a deformed (and stressed) configuration K may be decomposed into two deformations, such that

$$(10.5.1) \qquad x^k_{;K} = \Phi^k_{(\lambda)}\Theta^{(\lambda)}_K , \quad X^K_{;k} = \Theta^K_{(\lambda)}\Phi^{(\lambda)}_K ,$$

where $\underset{\sim}{\Theta}_{(\lambda)}$ and $\underset{\sim}{\Theta}^{(\lambda)}$ represent reciprocal triads of vectors, as well as $\underset{\sim}{\Phi}_{(\lambda)}$ and $\underset{\sim}{\Phi}^{(\lambda)}$.

The linear differential forms

$$(10.5.2) \qquad du^\lambda = \Phi^{(\lambda)}_K dx^k \quad \text{and} \quad du^\lambda = \Theta^{(\lambda)}_K dX^K$$

are in general non integrable. The vectors $\Phi^{(\lambda)}$ represent elastic distorsions, and $\underset{\sim}{\Theta}^{(\lambda)}$ are plastic or thermal distorsions (the terminology depends on the applications; in the theory of dislocations these distorsions are plastic). The coordinates u^{λ} owing to the non-integrability of (10.5.2) may be interpreted as coordinates of points of a non-Euclidean, linearly connected space with the coefficients of connection (with respect to the systems of reference x^k and X^K)

$$\Gamma^k_{\ell m} = \Phi^k_{(\lambda)} \partial_\ell \Phi , \quad \Gamma^K_{LM} = \Theta^k_{(\lambda)} \partial_L \Theta^{(\lambda)}_M . \qquad (10.5.3)$$

In the following sections we shall consider two special cases. In the first case we assume that the internal energy $\mathcal{E}$ is a function of the elastic distorsions and their gradients (Stojanović [422]),

$$\mathcal{E} = \mathcal{E}(\Phi^{\ell}_{(\lambda)}, \Phi^{\ell}_{(\lambda),m}) \qquad (10.5.4)$$

and in the second case we assume that $\mathcal{E}$ is a function of distortions and director gradients (Stojanović [421]),

$$\mathcal{E} = \mathcal{E}(\Phi^{\ell}_{(\lambda)}, d^{(\mu)}_{m,n}) . \qquad (10.5.5)$$

In the first case the theory may be reduced to the theory of elastic materials of grade two, and in the second case to theory of elastic generalized Cosserat materials.

10.5a Elastic Materials of Grade Two

We consider the local Clausius-Duhem inequality (8.24) in the form

$$-\varrho\dot{\Psi} - \varrho\eta\dot{\theta} + t^{(ij)}d_{ij} - m^{ijk}w_{ij,k} + \frac{1}{\theta}\theta_{,k}q^k \geq 0 \tag{10.5a.1}$$

and we assume that the free energy function Ψ is a function of $\overset{\ell}{\Phi}_{(\lambda)}$, $\overset{\ell}{\Phi}_{(\lambda),m}$ and of the temperature θ . Using (4.1.17) and (4.1.19), the inequality (10.5a.1) may be written in the form

$$\begin{aligned}&\left(-\varrho\frac{\partial\Psi}{\partial\overset{\ell}{\Phi}_{(\lambda)}} + g_{i\ell}t^{(ij)}\overset{(\lambda)}{\Phi}_j - m_\ell^{\cdot jk}\overset{(\lambda)}{\Phi}_{j,k} - m_m^{\cdot jk}\overset{(\mu)}{\Phi}_j\overset{m}{\Phi}_{(\mu),\ell}\overset{(\lambda)}{\Phi}_k\right)\dot{\overset{\ell}{\Phi}}_{(\lambda)} + \\ &+\left(-\frac{\partial\Psi}{\partial\overset{\ell}{\Phi}_{(\lambda),k}} - m_\ell^{\cdot jk}\overset{(\lambda)}{\Phi}_j\right)\overline{\dot{\overset{\ell}{\Phi}}}_{(\lambda),k} - \varrho\left(\frac{\partial\Psi}{\partial\theta} + \eta\right)\dot{\theta} + \frac{\theta_{,k}q^k}{\theta} \geq 0 .\end{aligned} \tag{10.5a.2}$$

This inequality is to be satisfied for arbitrary variations of $\dot{\overset{\ell}{\Phi}}_{(\lambda)}$, $\overline{\dot{\overset{\ell}{\Phi}}}_{(\lambda),m}$ and $\dot{\theta}$ and it will be satisfied if

$$t^{(ij)} = \varrho g^{i\ell}\left(\frac{\partial\Psi}{\partial\overset{\ell}{\Phi}_{(\lambda)}}\overset{j}{\Phi}_{(\lambda)} + \frac{\partial\Psi}{\partial\overset{\ell}{\Phi}_{(\lambda),k}}\overset{j}{\Phi}_{(\lambda),k} - \frac{\partial\Psi}{\partial\overset{k}{\Phi}_{(\lambda),j}}\overset{k}{\Phi}_{(\lambda),\ell}\right), \tag{10.5a.3}$$

$$m^{ijk} = -\varrho g^{i\ell}\frac{\partial\Psi}{\partial\overset{\ell}{\Phi}_{(\lambda),k}}\overset{j}{\Phi}_{(\lambda)} , \tag{10.5a.4}$$

$$\eta = -\frac{\partial\Psi}{\partial\theta} . \tag{10.5a.5}$$

Remains the inequality

$$\frac{1}{\theta}\theta_{,k}q^k \geq 0 \qquad (10.5a.6)$$

which is to be satisfied by the heat-conduction law.

The relations (10.5a.3-5) represent the constitutive equations for elastic incompatible deformations. It is to be noted that the couple-stress tensor in (10.5a.4) is completely determined.

When distorsions degenerate into deformation gradients, we put

$$\Phi^{\ell}_{(\lambda)} = x^{\ell}_{;\lambda}, \quad X^L = X^{\lambda}\delta^L_{\lambda}$$

and

$$\Phi^{\ell}_{(\lambda),k} = x^{\ell}_{;\lambda\mu}X^{\mu}_{;k},$$

and the constitutive equations for $t^{(ij)}$ and m^{ijk} reduce directly to (10.19) and (10.21). The indeterminacy of the couple stress tensor appears as a consequence of the assumption made <u>a priori</u> that the compatibility conditions are satisfied.

Introducing the request that the free energy function is invariant under rigid motions, and the right-hand side of (10.5a.4) possesses the same symmetries as the left-hand side, we obtain the system of linear differential equations

$$\left[g^{i\ell}\left(\frac{\partial\Psi}{\partial\Phi^{\ell}_{(\lambda)}}\Phi^{j}_{(\lambda)} + \frac{\partial\Psi}{\partial\Phi^{\ell}_{(\lambda),k}}\Phi^{j}_{(\lambda),k} - \frac{\partial\Psi}{\partial\Phi^{k}_{(\lambda),j}}\Phi^{k}_{(\lambda),\ell}\right)\right]_{[ij]} = 0 \qquad (10\ 5a.7)$$

(10.5a.8) $$\left(g^{il}\frac{\partial \Psi}{\partial \Phi^{l}_{(\lambda),k}}\Phi^{j}_{(\lambda)}\right)_{(ij)} = 0 \ .$$

There are 21 independent equations (10.5a.7-8) with one unknown function and 36 independent variables. This system admits 36-21=15 independent integrals.

It might be easily verified by direct calculation that the following material tensors satisfy the system of differential equations considered,

(10.5a.9) $$C_{AB} = C_{BA} = g_{ab}\Phi^{a}_{(\alpha)}\Phi^{b}_{(\beta)}\Theta^{(\alpha)}_{A}\Theta^{(\beta)}_{B} = g_{ab}x^{a}_{;A}x^{b}_{;B} \ ,$$

(10.5a.10) $$D_{ABC} = -D_{BAC} = g_{ab}\Phi^{a}_{(\alpha)}\Phi^{b}_{(\beta),m}\Phi^{m}_{(\gamma)}\Theta^{(\alpha)}_{[A}\Theta^{(\beta)}_{B]}\Theta^{(\gamma)}_{C} \ ,$$

where $\underset{\sim}{\Theta}^{(\alpha)}$ are distorsions introduced in (10.5.1).

The function Ψ which satisfies the system of equations (10.5a.7-8) is an arbitrary function of 15 independent components of the tensors $\underset{\sim}{C}$ and $\underset{\sim}{D}$ and of temperature,

$$\Psi = \Psi(\underset{\sim}{C}, \underset{\sim}{D}, \theta) \ ,$$

and we finally obtain after some calculations the following set of constitutive equations,

(10.5a.11) $$t^{[ij]} = 2\varrho\left[\frac{\partial \Psi}{\partial C_{AB}}x^{i}_{;A}x^{j}_{;B} + 2\frac{\partial \Psi}{\partial D_{ABC}}(x^{i}_{;A}x^{j}_{;BC} - x^{i}_{;A}x^{j}_{;L}T^{L}_{CB})_{(ij)}\right] ,$$

$$m^{ijk} = 2\varrho \frac{\partial \Psi}{\partial D_{ABC}} x^i_{;A} x^j_{;B} x^k_{;C} \;, \tag{10.5a.12}$$

where

$$T^L_{CA} = \overset{(\lambda)}{\theta}_{A,C} \overset{L}{\theta}_{(\lambda)} \;. \tag{10.5a.13}$$

10.5b Generalized Elastic Cosserat Materials

To derive the constitutive equations of the generalized elastic Cosserat medium with incompatible deformations we shall consider the strain energy function in the form(10.5.5), and apply the principle of virtual work (Stojanović [432]). We shall restrict our attention to the static case since we are here interested in the constitutive equations, and the equations of motion are not affected by incompatibilities. We assume the principle of virtual work in the form

$$\delta E = A \;, \tag{10.5b.1}$$

where

$$E = \int \varrho \mathcal{E} \, dv \;, \tag{10.5b.2}$$

and

$$A = \int_v \varrho (f_i \delta x^i + g^i_{(\lambda)} \delta \overset{(\lambda)}{d}_i) dv + \oint_S (F_i \delta x^i + G^i_{(\lambda)} \delta \overset{(\lambda)}{d}_i) ds \;. \tag{10.5b.3}$$

We assume that δx^i and $\delta d_i^{(\lambda)}$ are independent variations; f_i is the external body force, $g^i_{(\lambda)}$ are external director forces, F_i and $G^i_{(\lambda)}$ are surface tractions on the bounding surface s of v.

From (10.5b.2) and (10.5.5) we have

$$\text{(10.5b.4)} \qquad \delta E = \int_v \rho\left(\frac{\partial \varepsilon}{\partial \Phi^i_{(\lambda)}}\delta\Phi^i_{(\lambda)} + \frac{\partial \varepsilon}{\partial d^{(\lambda)}_{i,j}}\delta d^{(\lambda)}_{i,j}\right)dv .$$

By Appendix (A5.10) the expression (10.5b.4) will become

$$\text{(10.5b.5)} \quad \delta E = \int_v \rho\left[\left(\frac{\partial \varepsilon}{\partial \Phi^l_{(\lambda)}} - \frac{\partial \varepsilon}{\partial d^{(\mu)}_{i,j}} d^{(\mu)}_{i,l}\Phi^{(\lambda)}_j\right)\delta\Phi^l_{(\lambda)} + \frac{\partial \varepsilon}{\partial d^{(\mu)}_{i,j}}(\delta d^{(\mu)}_i)_{,j}\right]dv .$$

Writing

$$\text{(10.5b.6)} \qquad \rho\left(\frac{\partial \varepsilon}{\partial \Phi^l_{(\lambda)}} - \frac{\partial \varepsilon}{\partial d^{(\mu)}_{i,j}} d^{(\mu)}_{i,l}\Phi^{(\lambda)}_j\right)\Phi^m_{(\lambda)} = t_l^{\cdot m} ,$$

$$\text{(10.5b.7)} \qquad \rho\frac{\partial \varepsilon}{\partial d^{(\mu)}_{i,j}} = h^{ij}_{(\mu)} ,$$

and applying the modified divergence theorem (Appendix, (A5.9)) to (10.5b.5) we obtain the expression for the variation of the internal energy in the suitable form,

$$\text{(10.5b.8)} \quad \delta E = -\int_v (t_{l,j}^{\cdot j}\,\delta x^l + h^{\cdot ij}_{(\mu),j}\,\delta d^{(\mu)}_i)dv + \oint_S (t_l^{\cdot j}\delta x^l + h^{ij}_{(\mu)}\delta d^{(\mu)}_i)ds_j .$$

The principle of virtual work gives now the equilibrium equations

$$\text{(10.5b.9a)} \qquad t^{\cdot j}_{i,j} + \rho f_i = 0$$

$$h_{(\mu),j}^{\cdot ij} + \varrho g_{(\mu)}^{i} = 0 \qquad (10.5b.9b)$$

and the conditions on the bounding surface S,

$$t_i^{\cdot j} n_j = F_i \, ,$$

$$h_{(\mu)}^{\cdot ij} n_j = G_{(\mu)}^{i} \, . \qquad (10.5b.10)$$

The equations (10.5b.6–7) represent the constitutive equations, where $\underset{\sim}{t}$ is the stress tensor, and $\underset{\sim}{h}_{(\mu)}$ are three director stresses. The equation $(10.5b.10)_2$ is equivalent to (10.9), and $(10.5b.10)_1$ reduces to (10.7) when the distorsions degenerate into deformation gradients (and for $\dfrac{\partial \mathcal{E}}{\partial x^{l}_{;KL}} = 0$).

10.6 Thermoelasticity

Thermal deformations represent the best known example of incompatible deformations. If we denote by α the coefficient of thermal dilatation and by $\Theta(\underset{\sim}{X})$ the increment of temperature from an initially and everywhere in the body considered constant reference temperature T_0 =const., the strain tensor (in Cartesian coordinates)

$$e_{ij} = \alpha \Theta \delta_{ij} \qquad (10.6.1)$$

will not satisfy the compatibility conditions (4.11), unless the temperature Θ is constant, or a linear function of position coordinates.

To obtain the stress-strain relations in thermo-elasticity, we shall consider the distorsions $\overset{(\lambda)}{\theta}_L$, introduced in the section 4, as thermal distorsions. We further assume that thermal stresses are produced by the elastic distorsions $\overset{\ell}{\Phi}_{(\lambda)}$. For isotropic materials the thermal distorsions are isotropic functions, and for Cartesian coordinates we may write

$$\overset{(\lambda)}{\theta}_L = \vartheta(\underset{\sim}{X}, \theta)\delta^{\lambda}_{L} . \tag{10.6.2}$$

In this case T^{L}_{CA}, given by (10.5a.13), becomes

$$T^{L}_{CA} = \vartheta\vartheta_{,C}\delta^{L}_{A} , \tag{10.6.3}$$

and we have

$$D_{ABC} = C_{C[A,B]} = 2E_{C[A,B]} . \tag{10.6.4}$$

For isotropic materials the free energy Ψ is an isotropic function i.e. it is a function of isotropic invariants of the tensors $\underset{\sim}{C}, \underset{\sim}{D}$ and of the temperature (cf. Appendix, Sect. A2).

The constitutive equations (10.5a.11-12) reduce for isotropic materials to

$$t^{(ij)} = \varrho\left(\frac{\partial\Psi}{\partial E^{P}_{A}} G^{PB} x^{i}_{;A} x^{j}_{;B} + 2\frac{\partial\Psi}{\partial D^{P}_{.C}} \varepsilon^{PAB} x^{i}_{;A} x^{j}_{;BC}\right)_{(ij)} , \tag{10.6.5}$$

$$m_{l}^{\cdot k} = \varrho_0 \frac{\partial \Psi}{\partial D^{P}_{.C}} x^{k}_{;C} X^{P}_{;l} , \qquad (10.6.6)$$

where

$$\varrho_0 = \mathfrak{J}\varrho = \sqrt{\frac{g}{G}} \det X^{K}_{;k} , \quad D^{P}_{.C} = \frac{1}{2}\varepsilon^{PAB} D_{ABC} . \qquad (10.6.7)$$

To obtain linear constitutive equations it is sufficient to approximate Ψ by a polynomial quadratic in the strains,

$$\varrho_0 \Psi = A_1 I_E^2 + A_2 \overline{II}_E + A_3 I_E \Theta + A_4 II'_D + A_5 II''_D + A_6 \Theta^2 + \ldots \qquad (10.6.8)$$

where

$$I_E = E^{P}_{P}, \quad \overline{II}_E = E^{P}_{Q} E^{Q}_{P}, \quad II'_D = D^{P}_{.Q} D^{Q}_{.P}, \quad II''_D = D^{P}_{.Q} D_{P}^{.Q} . \qquad (10.6.9)$$

For infinitesimal deformation gradients, and for sufficiently small temperatures Θ we may write $\underset{\sim}{E} \approx \underset{\sim}{e}$ and the constitutive equations (10.6.5–6) become

$$t^{(ij)} = (2A_1 I_e + A_3 \Theta)\delta^{ij} + 2A_2 e^{ij} ,$$

$$m_{l}^{\cdot k} = 2A_4 D^{k}_{.l} + 2A_5 D_{l}^{\cdot k} . \qquad (10.6.10)$$

For the material constants $A_1, \ldots, A_5$ we may introduce the traditional notation,

$$2A_1 = \lambda , \quad A_3 = \alpha , \quad A_2 = G , \qquad (10.6.11a)$$

(10.6.11b) $A_4 = 2G\ell^2 , \quad A_5 = 2G\eta\ell^2 ,$

(cf. 10.2.19), and the equations (10.6.10) obtain the form in which they are well known in the linear theory of thermoelasticity with couple-stresses. These equations were first derived directly, within the frames of a linear theory by Nowacki [334]. Nowacki [333-338] developed the linear theory of the non-symmetric stress in thermoelasticity, without referring to the incompatibilities of the thermal strains, which is not necessary in linear theories. He derived the constitutive relations for both the materials of grade two, and for the Cosserat (i.e. micropolar) materials. Thermoelasticity of materials with microstructure, also without entering into the problems of incompatibilities, was studied by Wozniak in a number of papers [500-504, 506, 507] .

10.7 Dislocations

Dislocations are a kind of defects in the structure of matter. In the atomic structure of solids we can observe that the lattice points in real crystals are not perfectly arranged. A perfect arrangement of lattice points exists only in ideal crystals. In a real crystal, when compared with the corresponding perfect pattern, it is possible to observe vacant lattice points, atoms on the places where should not be an atom, extra atoms etc. Such defects are called by solid state physicists point defects. For mechanical properties of solids, primarily of metals, of greater importance are defects distributed on a surface which is bounded by a closed contour. For instance, all lattice points on a crystalographic plane bounded by a closed curve may be missing, or it is possible to have on this plane extra lattice points. Such two-dimensional defects are called dislocations. The curve bounding the surface upon which the missing or extra lattice points are located is the dislocation line, and this curve cannot be an open curve.

Crystals with dislocations may be compared with ideal crystals of the same crystalographic class. In the regions sufficiently far from the dislocation we say that the crystal is "good". A closed curve which encircles the dislocation line, passing through the lattice points in the "good" region of the crystal, is called the Burgers circuit. When a real crystal is compared with the (imagined) ideal crystal and when the Burgers circuit is mapped upon the ideal crystal, lattice point by lattice point, the curve in the ideal crystal will not be closed. The vector which measures this closure failure is called Burgers vector $\underset{\sim}{b}$. A dislocation is completely characterized by its dislacation line and by its Burgers vector.

Dislocations produce internal stresses in solids and these stresses cannot be associated to a uniquely defined field of displacements, i.e. the strain tensor which corresponds through the elastic stress-strain relations to the internal stresses produced by dislocations do not satisfy the compatibility conditions. The only way to release a body from internal stresses is to cut it.

Let us consider a body with an isolated dislocation, and let us consider a part of that body with the rectilinear segment of the dislocation line. The dislocation line can be isolated by a circular cylinder with a very small diameter. If we cut this element along a plane which is passing through the dislocation line, but with the cut ending on the cylinder, the element of the body will deform in order to release the internal stresses. Two portions of the body, facing one another along the plane of the cutting will suffer a displacement relative to one another. The displacements $\delta \underset{\sim}{u}$ of points, with the position vector $\underset{\sim}{r}$ with respect to an origin on the dislocation line, are given by the formula

$$\delta \underset{\sim}{u} = \underset{\sim}{b} + \underset{\sim}{d} \times \underset{\sim}{r}$$

$$\underset{\sim}{b} = \text{const.}, \quad \underset{\sim}{d} = \text{const}.$$

(Weingarten's theorem), where $\underset{\sim}{b}$ is the Burgers vector and $\underset{\sim}{d}$ is the rotation vector.

If we introduce a system of rectangular Cartesian coordinates, with the Z-axis along the dislocation line, the following classification of dislocations is due to Volterra. For $\underset{\sim}{d} = 0$ and $\underset{\sim}{b}$ parallel to one of the coordinate axes, X, Y and

Z respectively, the dislocations are of the 1st, 2nd or 3rd kind respectively, and for $\underset{\sim}{b}=0$ and $\underset{\sim}{d}$ parallel to one of the axes X, Y or Z, the dislocations are of the 4th, 5th or 6th kind, respectively. The dislocations usually considered in the literature on dislocations are belonging to the first three kinds of of Volterra dislocations. An arbitrary dislocation, in fact, has a constant Burgers vector, but its inclination to the dislocation line is changing along the line. The dislocations of the last three kinds are called sometimes <u>disclinations</u>.

a) Dislocations and Deformations of Directors

Let us regard simultaneously a crystal with dislocations and the corresponding perfect reference lattice. The lattice vectors $D^{(\lambda)}$ of the perfect crystal are determined by the lattice points and if the crystal is subjected to a deformation, the lattice vectors are deformed as material vectors. Hence, the lattice vectors of a perfect crystal cannot be considered as directors of a Cosserat medium. The lattice vectors in the perfect undeformed crystal represent fields of parallel vectors in the Euclidean sense.

If we refer the reference lattice to a coordinate system X^K, and the dislocated lattice to a coordinate system X^K, it is impossible to determine the lattice points of the dislocated crystal by the mappings of the form

$$x^k = x^k(\underset{\sim}{x}) \tag{10.7.1}$$

and the lattice vectors $\overset{(\lambda)}{\underset{\sim}{d}}$ of the dislocated crystal cannot be regarded as deformed lattice vectors $\overset{(\lambda)}{\underset{\sim}{D}}$ of the reference crystal i.e., there are no relations of the form

$$\overset{(\lambda)}{d}_k = \overset{(\lambda)}{D}_K X^K_{;k} . \tag{10.7.2}$$

If P is a lattice point of the dislocated crystal and if $\overset{(\lambda)}{D}_i$ are components of the lattice vectors of the reference crystal transported parallel to P, for the components of the lattice vectors $\overset{(\lambda)}{d}_i$ we may write

$$\overset{(\lambda)}{d}_i = \overset{(\lambda)}{D}_i + \overset{(\lambda)}{\Delta}_i . \tag{10.7.3}$$

The vectors $\overset{(\lambda)}{\underset{\sim}{\Delta}}$ vanish if the directors $\overset{(\lambda)}{d}_i$ deform as material vectors.

An infinitesimal displacement along the lattice vector $\overset{(\lambda)}{\underset{\sim}{d}}$ is represented by the expression

$$dr^\lambda = \overset{(\lambda)}{d}_i dx^i . \tag{10.7.4}$$

Let ℓ be a closed contour passing over lattice points in the "good "region of a dislocated crystal and surrounding a dislocation line (or zone with dislocations). The contour integral

$$\Delta b^{(\lambda)} = \oint_\ell dr^\lambda = \oint_\ell (\overset{(\lambda)}{D}_i + \overset{(\lambda)}{\Delta}_i) dx^i \tag{10.7.5}$$

determines the components of the Burgers vector in the directions of the lattice vectors $\overset{(\lambda)}{\underset{\sim}{d}}$. The Burgers vectors $\Delta \underset{\sim}{b}$ corresponding to the dislocations surrounded by ℓ is given by the components

$$\Delta b^{i} = \Delta b^{\lambda} d^{i}_{(\lambda)} \tag{10.7.6}$$

where $d^{i}_{(\lambda)}$ are vectors of the reciprocal director triad, $d^{i}_{(\lambda)} d^{(\lambda)}_{j} = \delta^{i}_{j}$.

For an infinitesimal region ΔF encircled by ℓ we have from (10.7.5)

$$\begin{aligned} \Delta b^{\lambda} &= \iint_{\Delta F} (D^{(\lambda)}_{[j,i]} + \Delta^{(\lambda)}_{[j,i]}) dF^{ij} \\ &= (D^{(\lambda)}_{[j,i]} + \Delta^{(\lambda)}_{[j,i]}) \Delta F^{ij} . \end{aligned} \tag{10.7.7}$$

Since the vectors $D^{(\lambda)}_{i}$ represent fields of parallel vectors, the gradients $D^{(\lambda)}_{j,i}$ vanish and we have

$$\Delta b^{\lambda} = \Delta^{(\lambda)}_{[j,i]} \Delta F^{ij} . \tag{10.7.8}$$

When $\Delta \underset{\sim}{F} \to 0$, we obtain from (10.7.6) and (10.7.8) for the disloca- tion density tensor $\alpha_{ij}^{\cdot\cdot k}$ the expression

$$\alpha_{ij}^{\cdot\cdot k} = d^{k}_{(\lambda)} \lim_{\Delta F \to 0} \frac{\Delta b^{\lambda}}{\Delta F^{ij}} = d^{k}_{(\lambda)} \Delta^{(\lambda)}_{[j,i]} . \tag{10.7.9}$$

(cf. Stojanović [419], and also Toupin [464]).

This relation, or its equivalent

$$\alpha_{ij}^{\cdot\cdot k} = d^{k}_{(\lambda)} d^{(\lambda)}_{[j,i]} = b^{k\ell} \alpha_{\ell j i} \tag{10.7.10}$$

where the fundamental metric tensor $\underset{\sim}{b}$ of the Euclidean space is used for the raising and lowering of indices, represents the basic relation between the distribution of dislocations and the gradients of directors [425, 419] .

The existence of the directors $\underset{\sim}{d}^{(\lambda)}$ for a given distribution of dislocations depends on the integrability of the equations (10.7.10), which we can write in the form

(10.7.11) $$\partial_i d_j^{(\lambda)} - \partial_j d_i^{(\lambda)} = 2\alpha_{ij}^{\cdot\cdot t} d_t^{(\lambda)} .$$

Differentiating this relation with respect to x^k and alternating the indices ijk we obtain

(10.7.12) $$\partial_{[k}\partial_i d_{j]}^{(\lambda)} = d_t^{(\lambda)}\partial_{[k}\alpha_{ij]}^{\cdot\cdot t} + \alpha_{[ij}^{\cdot\cdot t}\partial_{k]} d_t^{(\lambda)} .$$

The left-hand side of (10.7.12) vanishes because of the commutativity of partial derivatives, and the integrability conditions reduce to the relations

(10.7.13) $$\partial_{[k}\alpha_{ij]}^{\cdot\cdot \ell} = -[d_{(\lambda)}^{\ell}(\partial_k d_t^{(\lambda)})\alpha_{ij}^{\cdot\cdot t}]_{[ijk]} .$$

The indices ijk involved in the alternation in (10.7.13) must all have different values and hence there are only three independent relations (10.7.13) for ℓ =1,2,3. Nothing will be lost if we transvect the relations with the alternating Ricci Tensor $\underset{b}{\varepsilon}^{ijk}$ formed with respect to the Euclidean metric tensor $\underset{\sim}{b}$. Writing

(10.7.14) $$\frac{1}{2}\underset{b}{\varepsilon}^{ijk}\alpha_{ij}^{\cdot\cdot\ell} = \alpha^{k\ell} ,$$

and

$$d^{\ell}_{(\lambda)}\partial_k d^{(\lambda)}_t = -d^{(\lambda)}_t \partial_k d^{\ell}_{(\lambda)} = D^{\ell}_{kt} \qquad (10.7.15)$$

the integrability conditions (10.7.13) obtain the form

$$\partial_k \alpha^{k\ell} + b^{k}_{km}\alpha^{m\ell} = -D^{\ell}_{mt}\alpha^{mt} . \qquad (10.7.16)$$

Here b^{m}_{kt} are the Christoffel symbols of the first kind for the tensor $\underset{\sim}{b}$ and $b^{m}_{km} = \partial_k \ln\sqrt{b}$.

b) Geometry

In the continuum theory of dislocations the stress-free state (N) of a dislocated crystal is considered in a linearly connected metric space with torsion [247] . If g_{ij} is the fundamental tensor of this space and $S_{ij}^{\cdot\cdot k}$ the torsion tensor, the coefficients of connection Γ^{k}_{ij} are given by

$$\Gamma^{k}_{ij} = g^{\,\cdot k}_{ij} + h_{ij}^{\cdot\cdot k} , \qquad (10.7.17)$$

where $g^{\,\cdot k}_{ij}$ are the Christoffel symbols of the second kind for the tensor $\underset{\sim}{g}$ and

$$\begin{aligned} h_{ij}^{\cdot\cdot k} &= S_{ij}^{\cdot\cdot k} - S_{j\cdot i}^{\cdot k} + S^{k}_{\cdot ij} , \\ S_{ij}^{\cdot\cdot k} &= \Gamma^{k}_{[ij]} . \end{aligned} \qquad (10.7.18)$$

Writing

$$g_{ijk} = \frac{1}{2}\left(\overset{b}{\nabla}_i g_{jk} + \overset{b}{\nabla}_j g_{ki} - \overset{b}{\nabla}_k g_{ij}\right) , \qquad (10.7.19)$$

where $\overset{b}{\nabla}_m$ denotes the covariant differentiation with respect to the Euclidean metric tensor $\underset{\sim}{b}$ [247], the coefficients Γ_{ij}^{k} may be expressed by the relations

$$(10.7.20) \qquad \Gamma_{ij}^{k} = b_{ij}^{k} + g^{kl} g_{ijl} + h_{ij}^{\cdot\cdot k} \equiv b_{ij}^{k} + G_{ij}^{\cdot\cdot k} .$$

If we assume that the lattice vectors of a dislocated crystal represent fields of parallel vectors in the space L_3, they have to be covariant constant with respect to the connection Γ_{ij}^{k},

$$(10.7.21) \qquad \overset{\Gamma}{\nabla}_i d_j^{(\lambda)} = \partial_i d_j^{(\lambda)} - \Gamma_{ij}^{k} d_k^{(\lambda)} + 0 ,$$

and from (10.7.15) it follows that

$$(10.7.22) \qquad \Gamma_{ij}^{k} = D_{ij}^{k} = d_{(\lambda)}^{k} \partial_i d_j^{(\lambda)} .$$

Hence, the geometry of the non-Euclidean space L_3 is completely determined by the directors $d^{(\lambda)}$, i.e. by the lattice vectors of the dislocated crystal.

From (10.7.10) and (10.7.22) we see that the torsion tensor $S_{ij}^{\cdot\cdot k}$ of L_3 is equal to the dislocation density tensor,

$$(10.7.23) \qquad S_{ij}^{\cdot\cdot k} = \alpha_{ij}^{\cdot\cdot k} .$$

The integrability condition (10.7.16) may be brought to a more familiar form. If we substitute partial derivatives by the covariant derivatives with respect to the Euclidean metric $\underset{\sim}{b}$, i.e.

$$\partial_k \alpha^{kl} \equiv \overset{b}{\nabla}_k \alpha^{kl} - b^{\,k}_{km} \alpha^{ml} - b^{\,l}_{km} \alpha^{km} , \qquad (10.7.24)$$

and if we use the expression (10.7.20) for the coefficients of connection, the expression (10.7.16) reduces to

$$\overset{b}{\nabla}_k \alpha^{kl} = -G_{km}^{\cdot\cdot l} \alpha^{km} . \qquad (10.7.25)$$

Using the fundamental tensor g_{ij} of L_3 for the raising and lowering of the indices, so that

$$\alpha^{kl} g_{lj} = \alpha^{k}_{\cdot j} , \qquad (10.7.26)$$

the integrability conditions obtain the form

$$\overset{b}{\nabla}_k \alpha^{k}_{\cdot j} = g^{kl} G_{ijk} \alpha^{il} . \qquad (10.7.27)$$

This coincides with Kröner's and Seeger's generalization to the non-linear case of the conservation law for the dislocation density tensor, given in the linear theory by Nye.

In the treatment of the continuously distributed dislocations Kondo and Kröner and Seeger [247, 248] consider the space L_3 corresponding to the (N)-configuration of a dislocated crystal with the coefficients of connection determined in terms of the distorsions $\Phi^{(\lambda)}_{l}$,

$$\tilde{\Gamma}^{k}_{lm} = \Phi^{k}_{(\lambda)} \partial_l \Phi^{(\lambda)}_{m} . \qquad (10.7.28)$$

The coefficients Γ^{k}_{lm} determined in terms of the directors $\underset{\sim}{d}_{(\lambda)}$ were introduced first by Bilby et al. However, the geometries of the two spaces, L_3 and $\tilde{L}_3$ are equivalent. In L_3 the dislocation density tensor is also equal to the torsion tensor of the space,

$$\alpha_{\ell m}{}^{k} = \tilde{\Gamma}^{k}_{[\ell m]} = \Phi^{k}_{(\lambda)} \partial_{[\ell} \Phi^{(\lambda)}_{m]} . \tag{10.7.29}$$

The integrability condition of (10.7.29) reads

$$\overset{b}{\nabla}_{k} \alpha^{k\ell} + b^{k}_{km} \alpha^{m\ell} = -\tilde{\Gamma}^{\ell}_{mt} \alpha^{mt} . \tag{10.7.30}$$

Comparing this with (10.7.16) we see that the coefficients of connection Γ^{k}_{ij} and $\tilde{\Gamma}^{k}_{ij}$ of the spaces L_3 and $\tilde{L}_3$ are equal, which makes the geometries equivalent.

The time does not permit us here to discuss the problem of internal stresses, but we shall note here that the theory of internal stresses contributed very much to the increase of interest in incompatible deformations and in the theory of elasticity with the non-symmetric stress tensor. (Cf. Kröner [252, 253, 255, 256]). Hehl and Kröner have calculated directly couple-stresses for an isolated dislocation [200] . An increasing number of papers deals now with dislocations in directed media. Claus and Eringen [62] approached this problem from the point of view of micromorphic mechanics and gave a comparative analysis of some other contributions in this field. Cf. also Ben-Abraham [29] , Minagawa [281] and Claus and Eringen [62] .

The linear theory of moving dislocations in the Cosserat continuum was also recently treated by Schäfer [317-319] and by Kluge [240].

c) Disclinations

One type of disclinations, which corresponds to Volterra dislocations of the sixth kind, called wedge disclinations, has been detected experimentally in the two-dimensional lattice formed by vortex lines in the mixed state of type II superconductors.

Since the disclinations represent a rotational closure failure, in analogy to dislocations, they can be associated to the incompatibilities of rotation of a Cosserat triad of directors.

According to (10.28), the compatibility conditions for the director deformation read

$$F_{\alpha[AB,C]} = 0 \tag{10.7.31}$$

and for infinitesimal rotations this reduces to

$$x^{m}_{\cdot[\ell,n]} = 0$$

which may be also written in the form

$$\theta^{ij} = \varepsilon^{j\ell n} x^{i}_{\cdot\ell,n} = 0 .$$

If these compatibility conditions are not satisfied, the tensor θ^{ij} represents the disclination density tensor (Anthony, Essmann, Seeger and Träuble [13] , Claus and Eringen

[62] . Up till now the theory is not much developed.

11. Shells Plates and Rods

We mentioned already that in the theories of thin bodies, with one (or two) dimensions small in comparison with the remaining two (or one) dimensions of the body, the equations valid for the three-dimensional continuum may be simplified. This is of the greatest technical importance. Different approximations of the three-dimensional equations lead to different models, but the common characteristic of all these models is that the orientation of the elements, the presence of couple-stresses and hyperstresses etc. appear as a result of the approximation and as a substitute of the neglected thickness of the body considered.

In 1958 Ericksen and Truesdell [121] gave an analysis of stress and strain in rods and shells from the point of view of the theory of oriented bodies, and they indicated the significance of couple-stresses in the exact description of the state of stress. Their considerations were based on the geometry of rods and shells and they have not made any constitutive assumptions.

Since 1958 a large number of papers appeared, mostly dealing with elastic shells and rods. In this section we shall give only a brief review of some of the most characteristic approaches to this important part of Applied Mechanics. Our attention will be concentrated on the theory of shells with only a very short account of some of the ideas which appeared recently. We refer here also to the references quoted at the end of the sections 7.2 and 7.3.

11.1a Theories with Rigid Directors

In 1958 Günther [189] considered the Cosserat continuum with rigid director triads and assumed that the points of the continuum have six degrees of freedom, so that at each point we may consider a displacement vector $\underset{\sim}{u}$ and a rotation vector $\underset{\sim}{\Phi}$ which is independent of $\underset{\sim}{u}$. The deformation is determined by the deformation vectors

$$\underset{\sim}{\varepsilon}_i = \partial_i \underset{\sim}{u} + \underset{\sim}{g}_i \times \underset{\sim}{\Phi} , \tag{11.1a.1}$$

$$\underset{\sim}{x}_i = \partial_i \underset{\sim}{\Phi} , \tag{11.1a.2}$$

with the components

$$\varepsilon_{ij} = u_{j,i} - \varepsilon_{ijk}\Phi^k , \tag{11.1a.3}$$

$$x_i^{\cdot \ell} = \Phi^{\ell}_{,i} . \tag{11.1a.4}$$

Thus we see that the kinematics of Günther coincides with the kinematics of micropolar media (cf. section (10.4). The symmetric part of ε_{ij} corresponds to the strain tensor $e_{ij} = \varepsilon_{(ij)}$ of the linear theory, and the antisymmetric part represents what might be called a resultant rotation, composed of the rotation induced by the displacement and of an independent rotation $\underset{\sim}{\Phi}$,

$$\text{(11.1a.5)} \qquad \varepsilon_{[ij]} = w_{ji} - \varepsilon_{ijk}\Phi^{k} .$$

The static equations may be obtained from the principle of virtual work. Let $\underset{\sim}{f}$ be the volume force and $\underset{\sim}{l}$ the volume couple acting on points of the body v and $\underset{\sim}{q}$ and $\underset{\sim}{p}$ the surface tractions and couples acting on the body surface S bounding v. All forces and couples are in equilibrium if

$$\text{(11.1a.6)} \qquad \int_{v} \varrho(\underset{\sim}{f}\cdot\delta\underset{\sim}{u} + \underset{\sim}{l}\cdot\delta\underset{\sim}{\Phi})dv + \oint_{S}(\underset{\sim}{q}\cdot\delta\underset{\sim}{u} + \underset{\sim}{p}\cdot\delta\underset{\sim}{\Phi})ds = 0 .$$

For rigid motions

$$\text{(11.1a.7)} \qquad \begin{aligned} \delta\underset{\sim}{\varepsilon}_{i} &= \partial_{i}(\delta\underset{\sim}{u}) + \underset{\sim}{g}_{i}\times\delta\underset{\sim}{\Phi} = 0 , \\ \delta x_{i} &= \partial_{i}(\delta\underset{\sim}{\Phi}) = 0 . \end{aligned}$$

Multiplying (11.1a. 7) by Langragian multipliers $t^{i}dv$ and $m^{i}ds$, respectively, integrating the first of these relations over v and the other over s and subtracting the so obtained expressions from (11.1a.6) we obtain

$$\int_v (\underset{\sim}{t}^i \cdot \delta \underset{\sim}{\varepsilon}_i + \underset{\sim}{m}^i \cdot \delta \underset{\sim}{x}_i - \varrho \underset{\sim}{f} \cdot \delta \underset{\sim}{u} - \varrho \underset{\sim}{l} \cdot \delta \underset{\sim}{\Phi}) dv - \oint_s (\underset{\sim}{q} \cdot \delta \underset{\sim}{u} + \underset{\sim}{p} \cdot \delta \underset{\sim}{\Phi}) ds = 0 . \tag{10.11a.8}$$

For arbitrary $\delta \underset{\sim}{u}$ and $\delta \underset{\sim}{\Phi}$ follow now the equations which correspond to (7.37) and (7.41) for $\underset{\sim}{a} = 0$, $\dot{\underset{\sim}{\sigma}} = 0$, and the boundary conditions

$$\underset{\sim}{t}^i n_i = \underset{\sim}{q} , \quad \underset{\sim}{m}^i n_i = \underset{\sim}{p} , \tag{10.11a.9}$$

where $\underset{\sim}{n}$ is the unit normal to s .

This approach to the mechanics of Cosserat continua Günther applied in 1961 [190] to the theory of shells. Let σ be the middle surface of a shell, and $\underset{\sim}{n}$ the unit normal to σ . If x^α, $\alpha = 1,2$ are coordinates on σ, the rotation and displacement vectors for the points on σ are given by

$$\underset{\sim}{\Phi} = \Phi^\alpha \underset{\sim}{a}_\alpha + \Phi \underset{\sim}{n} , \quad \underset{\sim}{u} = u_\alpha \underset{\sim}{a}^\alpha + u \underset{\sim}{n} , \tag{10.11a.10}$$

where $\underset{\sim}{r} = \underset{\sim}{r}(x^1, x^2)$ is the position vector for points of the middle surface, and $\underset{\sim}{a}_\alpha$ are the base vectors defined by the relations

$$\underset{\sim}{a}_\alpha = \frac{\partial \underset{\sim}{r}}{\partial x^\alpha} , \quad \underset{\sim}{g}_3 = \underset{\sim}{n}(x^1, x^2) .$$

The deformation vectors are

(11.1a.11) $$\underset{\sim}{\chi}_\alpha = \partial_\alpha \underset{\sim}{\Phi} , \quad \underset{\sim}{\varepsilon}_\alpha = \partial_\alpha \underset{\sim}{u} + \underset{\sim}{g}_\alpha \times \underset{\sim}{\Phi} ,$$

or

(11.1a.12) $$\underset{\sim}{\chi}_\alpha = \chi_\alpha^{\cdot\beta} \underset{\sim}{g}_\beta + \chi_\alpha \underset{\sim}{n} , \quad \underset{\sim}{\varepsilon}_\alpha = \varepsilon_{\alpha\beta} \underset{\sim}{g}^\beta + \varepsilon_\alpha \underset{\sim}{n}$$

with the components

(11.1a.13)
$$\chi_\alpha^{\cdot\beta} = \Phi^\beta_{,\alpha} - b_\alpha^\beta \Phi , \quad \chi_\alpha = \Phi_{,\alpha} + b_{\alpha\beta}\Phi^\beta ,$$
$$\varepsilon_{\alpha\beta} = u_{\beta,\alpha} - b_{\alpha\beta} u - e_{\alpha\beta}\Phi , \quad \varepsilon_\alpha = u_{,\alpha} + b_\alpha^\beta u_\beta + e_{\alpha\beta}\Phi^\beta .$$

Here we used the notation

(11.1a.14) $$\underset{\sim}{e}_\alpha = \underset{\sim}{n} \times \underset{\sim}{a} , \quad b_\alpha = -\partial_\alpha \underset{\sim}{n} , \quad a_{\alpha\beta} = \underset{\sim}{a}_\alpha \cdot \underset{\sim}{g}_\beta ,$$

where $b_{\alpha\beta}$ is the second fundamental tensor of the surface,

(11.1a.15)
$$b_{\alpha\beta} = \underset{\sim}{b}_\alpha \cdot \underset{\sim}{a}_\beta = -\underset{\sim}{a}_\beta \cdot \partial_\alpha \underset{\sim}{n} = b_{\beta\alpha} ,$$
$$b_\alpha^\beta = \underset{\sim}{b}_\alpha \cdot \underset{\sim}{a}^\beta = a^{\beta\gamma} b_{\alpha\gamma} ,$$

and $\varepsilon_{\alpha\beta}$ is the two-dimensional permutation (Ricci) tensor

(11.1a.16)
$$\epsilon_{\alpha\beta} = \underset{\sim}{n}(\underset{\sim}{a}_\alpha \times \underset{\sim}{a}_\beta) = -\epsilon_{\beta\alpha} = \sqrt{a}\, e_{\alpha\beta} ,$$
$$(e_{11} = e_{22} = 0 , \quad e_{12} = -e_{21} = 1) .$$

Günther introduced certain "response quantities" $\underset{\sim}{K}$ and $\underset{\sim}{M}$ of the shell, defined by

$$\underset{\sim}{K} = \underset{\sim}{K}^{\beta}\nu_{\beta} = (K^{\alpha\beta}\underset{\sim}{a}_{\alpha} + K^{\beta}\underset{\sim}{n})\nu_{\beta} \,,$$
$$\underset{\sim}{M} = \underset{\sim}{M}^{\beta}\nu_{\beta} = (\sqrt{a}M^{\alpha\beta}\underset{\sim}{a}_{\alpha} + \sqrt{a}M^{\beta}\underset{\sim}{n})\nu_{\beta} \,, \qquad (11.1a.17)$$

where $\underset{\sim}{\nu}$ is the unit normal to an arbitrary closed curve C in the middle surface σ, and $\underset{\sim}{n}$ is the unit normal to σ. Let $\underset{\sim}{f}$ and $\underset{\sim}{\ell}$ be external force and couple acting on the points of the middle surface, and let $d\underset{\sim}{K}$ and $d\underset{\sim}{M}$ be forces and couples acting on the points of the bounding curve C of σ . Günther postulated the principle of virtual work in the form

$$-\iint_{\sigma} e^{\alpha\beta}(\underset{\sim}{K}_{\alpha}\cdot\delta\underset{\sim}{\varepsilon}_{\beta} + \underset{\sim}{M}_{\alpha}\cdot\delta\underset{\sim}{\chi}_{\beta})d\sigma = \iint_{\sigma} \varrho(\underset{\sim}{f}\cdot\delta\underset{\sim}{u} + \underset{\sim}{\ell}\cdot\delta\underset{\sim}{\Phi})d\sigma +$$
$$+ \oint_{C}(d\underset{\sim}{K}\cdot\delta\underset{\sim}{u} + d\underset{\sim}{M}\cdot\delta\underset{\sim}{\Phi})dC \,. \qquad (11.1a.18)$$

Assuming that the vectors $\delta\underset{\sim}{u}$ and $\delta\underset{\sim}{\Phi}$ may be varied arbitrarily, introducing (11.1a.7) into (11.1a.8) and applying the divergence theorem we find the system of equilibrium equations for points on the middle surface,

$$\epsilon^{\alpha\beta}\partial_{\alpha}\underset{\sim}{K}_{\beta} + \varrho\underset{\sim}{f} = 0 \,,$$
$$\epsilon^{\alpha\beta}(\partial_{\alpha}\underset{\sim}{M}_{\beta} + \underset{\sim}{g}_{\alpha}\times\underset{\sim}{K}_{\beta}) + \varrho\underset{\sim}{\ell} = 0 \,, \qquad (11.1a.19)$$

and for the points on the bounding curve C,

$$\underset{\sim}{K}_\alpha dx^\alpha = d\underset{\sim}{K}, \quad \underset{\sim}{M}_\alpha dx^\alpha = d\underset{\sim}{M}. \tag{11.1a.20}$$

If we write now

$$e^{\alpha\beta}\underset{\sim}{K}_\beta = \sqrt{a}\underset{\sim}{N}^\alpha, \quad e^{\alpha\beta}K_\beta = \sqrt{a}N^\alpha, \tag{11.1a.21}$$

the quantities $N^{\beta\alpha}$ represent the components of the shell forces (membrane forces), and N^α are components of the transversal force.

Denoting again by " $|_\lambda$ " covariant differentiation with respect to the coordinates x^λ on σ, scalar multiplication of the equations (11.1a.19) with the base vectors will give the equilibrium equations in the componental form,

$$N^{\alpha\beta}{}_{|\beta} - b^\alpha_\beta N^\beta + \frac{g}{\sqrt{a}} f^\alpha = 0, \quad N^\alpha{}_{|\alpha} + b_{\alpha\beta}N^{\alpha\beta} + \frac{g}{\sqrt{a}} f^3 = 0; \tag{11.1a.22}$$

$$\begin{cases} e^{\alpha\beta}(M_{\nu\beta|\alpha} - b_{\alpha\nu}M_\beta) - \epsilon_{\alpha\nu}\sqrt{a}N^\alpha + g\ell_\nu = 0 \\ e^{\alpha\beta}(M_{\beta|\alpha} + b^\mu_\alpha M_{\mu\beta}) + \epsilon_{\alpha\beta}\sqrt{a}N^{\beta\alpha} + g\ell_3 = 0. \end{cases} \tag{11.1a.23}$$

Let $\underset{\sim}{m}^\beta = m^{\alpha\beta}\underset{\sim}{a}_\alpha + m^\beta\underset{\sim}{n}$ be some new moments, related to the moments $\underset{\sim}{M}^\beta$ by the relations

$$\underset{\sim}{m}^\beta = \epsilon^{\beta\gamma}[\underset{\sim}{M}_\gamma \times \underset{\sim}{n} + (\underset{\sim}{M}_\gamma \cdot \underset{\sim}{n})\underset{\sim}{n}],$$

or, in the componental form,

$$m^{\alpha\beta} = \epsilon^{\alpha\lambda}\epsilon^{\beta\mu}M_{\lambda\mu}, \quad m^{\beta} = \epsilon^{\beta\lambda}M_{\lambda}. \qquad (11.1a.24)$$

The equilibrium equations (11.1a.23) obtain now the form

$$\begin{aligned} m^{\alpha\beta}{}_{|\beta} - \sqrt{a}N^{\alpha} + \epsilon^{\alpha\beta}b_{\beta\gamma}m^{\gamma} + \varrho\epsilon^{\alpha\beta}\ell_{\beta} &= 0 \, , \\ m^{\alpha}{}_{|\alpha} + \epsilon_{\alpha\beta}(\sqrt{a}N^{\alpha\beta} + b^{\beta}_{\lambda}m^{\alpha\lambda}) + \varrho\ell_3 &= 0 \, . \end{aligned} \qquad (11.1a.25)$$

The equilibrium equations (11.1a.23) are essentially the same as the equilibrium equations (7.2.10) for Cosserat surfaces in the static case. The difference appears when we compare the equations (11.1a.25) with (7.2.12), since the later equations do not include the influence of the forces $\underset{\sim}{N}^{\alpha}$ upon the director stresses $\underset{\sim}{h}^{\alpha}$. This difference is a consequence of different kinematical models which served as bases of the theories.

To establish a connection between the forces and moments $\underset{\sim}{K}$, $\underset{\sim}{M}$ (or $\underset{\sim}{N}$ and $\underset{\sim}{m}$) acting on the points of the middle surface σ, and the usual three-dimensional stress tensor $\underset{\sim}{t}$, we shall assume that the position of points of the shell are determined by the coordinates x^{α},(α = 1,2) on σ and by the normal distance $x^3 = z$ of the points considered from σ. Thus, for an arbitrary point of the shell we may write

(11.1a.26) $$\underset{\sim}{r}^* = \underset{\sim}{r}(x^1, x^2) + z\underset{\sim}{n}(x^1, x^2) .$$

The base vectors $\underset{\sim}{g}_i$ at $\underset{\sim}{r}^*$ are

(11.1a.27)
$$\underset{\sim}{g}_\alpha = \frac{\partial \underset{\sim}{r}^*}{\partial x^\alpha} = \underset{\sim}{a}_\alpha + z\partial_\alpha \underset{\sim}{n} = \underset{\sim}{a}_\alpha - z\underset{\sim}{b}_\alpha ,$$
$$\underset{\sim}{g}_3 = \underset{\sim}{n} ,$$

and the components of the fundamental tensor g_{ij} are

(11.1a.28)
$$g_{\alpha\beta} = a_{\alpha\beta} - 2zb_{\alpha\beta} + z^2 \underset{\sim}{b}_\alpha \cdot \underset{\sim}{b}_\beta ,$$
$$g_{\alpha 3} = 0 ,$$
$$g_{33} = \underset{\sim}{n} \cdot \underset{\sim}{n} = 1 .$$

Considering the shell as a three-dimensional body we assume that the stress vector $\underset{\sim}{t}$ is defined for the surface elements orthogonal to the middle surface σ , i.e. $\underset{\sim}{t}^3 = 0$ and

(11.1a.29) $$\underset{\sim}{t} = t^{i\alpha} \underset{\sim}{g}_i \nu_\alpha ,$$

where ν_α are components of the unit normal to an arbitrary curve C on σ . If dC is the arc element of C with the unit normal $\underset{\sim}{\nu}$, the contact force $d\underset{\sim}{K}$ acting on the surface element $\nu dCdz$ will be

(11.1a.30) $$d\underset{\sim}{K} = \underset{\sim}{t}^\beta \nu_\beta dCdz .$$

Let us denote the unit tangent vector to C by $\underset{\sim}{\tau}$,

$$\underset{\sim}{\tau} = \frac{dx^\lambda}{dC}\underset{\sim}{g}_\lambda . \qquad (11.1a.31)$$

Then we have

$$\nu_\beta = (\underset{\sim}{\tau}\times\underset{\sim}{n})\cdot\underset{\sim}{g}_\beta = (\underset{\sim}{g}_\beta\times\underset{\sim}{g}_\lambda)\cdot\underset{\sim}{n}\frac{dx^\lambda}{dC} . \qquad (11.1a.32)$$

However, according to (11.1a.16) we may write

$$\begin{aligned} \underset{\sim}{g}_\beta\times\underset{\sim}{g}_\lambda &= \sqrt{g}\, e_{\beta\lambda} = \sqrt{\frac{g}{a}}\epsilon_{\beta\lambda} = h\epsilon_{\beta\lambda} , \\ g &\equiv g_{11}g_{22} - g_{12}g_{12} , \end{aligned} \qquad (11.1a.33)$$

and we have

$$d\underset{\sim}{K} = \epsilon_{\beta\lambda} h t^{i\beta}\underset{\sim}{g}_i dz dx^\lambda . \qquad (11.1a.34)$$

From $(11.1a.20)_1$ we see that along C

$$d\underset{\sim}{K} = d\underset{\sim}{K}_\lambda dx^\lambda ,$$

and therefore

$$d\underset{\sim}{K}_\lambda = \epsilon_{\beta\lambda} h t^{i\beta}\underset{\sim}{g}_i dz . \qquad (11.1a.35)$$

Introducing the "reduced stress tensor" $\sigma^{i\beta}$ by the relations

$$t^{i\beta}\underset{\sim}{g}_i \equiv \sigma^{i\beta}\underset{\sim}{a}_i = \sigma^{\alpha\beta}\underset{\sim}{a}_\alpha + \sigma^{3\beta}\underset{\sim}{n} , \qquad (11.1a.36a)$$

$$\sigma^{\alpha\beta} = t^{i\beta}\underset{\sim}{g}_i\cdot\underset{\sim}{a}^\alpha = t^{\lambda\beta}\underset{\sim}{g}_\lambda\cdot\underset{\sim}{a}^\alpha =$$

(11.1a.36b) $$= t^{\alpha\beta} - zb^\alpha_\lambda t^{\lambda\beta} ,$$

$$\sigma^{3\beta} = t^{3\beta} ,$$

we see that

(11.a.37) $$d\underset{\sim}{K}_\lambda = \epsilon_{\beta\lambda}(h\sigma^{\alpha\beta}\underset{\sim}{a}_\alpha + ht^{3\beta}\underset{\sim}{n})dz .$$

Integrating over the thickness $-\frac{a}{2} \leq z \leq \frac{a}{2}$ of the shell we obtain the forces $\underset{\sim}{K}_\lambda$,

(11.1a.38) $$\underset{\sim}{K}_\lambda = \epsilon_{\beta\lambda}\left(\int_{-\frac{a}{2}}^{\frac{a}{2}} h\sigma^{i\beta}dz\right)\underset{\sim}{a}_i .$$

By (11.1a.21) we find

(11.1a.39) $$\sqrt{a}\underset{\sim}{N}^\beta = \underset{\sim}{a}_i\int_{-\frac{a}{2}}^{\frac{a}{2}} h\sigma^{i\beta}dz ,$$

or

(11.1a.40) $$\sqrt{a}N^{\alpha\beta} = \int_{-\frac{a}{2}}^{\frac{a}{2}} h\sigma^{\alpha\beta}dz , \quad \sqrt{a}N^\beta = \int_{-\frac{a}{2}}^{\frac{a}{2}} ht^{3\beta}dz .$$

Günther also considered the moment

(11.1a.41) $$d\underset{\sim}{M} = z\underset{\sim}{n}\times d\underset{\sim}{K} = \epsilon_{\beta\lambda}hz\sigma^{i\beta}dz\,dx^\lambda\underset{\sim}{\varepsilon}_i ,$$

where

$$\underset{\sim}{\varepsilon}_i \equiv \underset{\sim}{n} \times \underset{\sim}{a}_i \ . \tag{11.1a.42}$$

Since $\underset{\sim}{\varepsilon}_3 = 0$, by $(11.1a.20)_2$ we find

$$d\underset{\sim}{M}_\lambda = \underset{\sim}{\varepsilon}_\alpha \epsilon_{\beta\lambda} hz\sigma^{\alpha\beta} dz \ , \tag{11.1a.43}$$

and

$$dM_{\mu\lambda} = \epsilon_{\alpha\mu}\epsilon_{\beta\lambda} hz\sigma^{\alpha\beta} dz \ . \tag{11.1a.44}$$

Integration over $-\frac{a}{2} \leq z \leq \frac{a}{2}$ gives

$$\begin{aligned} M_{\mu\lambda} &= \epsilon_{\alpha\mu}\epsilon_{\beta\lambda} \int_{-\frac{a}{2}}^{\frac{a}{2}} hz\sigma^{\alpha\beta} dz \ , \\ M_{3\lambda} &= 0 \ . \end{aligned} \tag{11.1a.45}$$

Comparing the results with (11.1a.24) we obtain the expressions for the moments induced by the stresses,

$$m^{\alpha\beta} = \int_{-\frac{a}{2}}^{\frac{a}{2}} hz\sigma^{\alpha\beta} dz \ , \quad m^\beta = 0 \ . \tag{11.1a.46}$$

Obviously the moments $m^{\alpha\beta}$ are directly connected with the stress field in the shell. When the three-dimensional theory is reduced to the two-dimensional theory, for a more complete picture of the stress-field it is necessary to consider not only the resultant forces $\underset{\sim}{N}_\lambda$,but also the resultant couples

$\underset{\sim}{m}$.

According to $(11.1a.46)_2$, the equilibrium equations $(11.1a.25)_2$ reduce to

$$(11.1a.47) \qquad \sqrt{a}\,\epsilon_{\alpha\beta}N^{\alpha\beta} + \epsilon_{\alpha\beta}b^{\beta}_{\lambda}m^{\alpha\lambda} + \sigma\ell_3 = 0 ,$$

which is equivalent to

$$(11.1a.48) \qquad \sqrt{a}(N^{12} - N^{21}) - m^{1\lambda}b^{2}_{\lambda} + m^{2\lambda}b^{1}_{3} + \underset{\sim}{g}\ell_3 = 0 .$$

From $(11.1a.23)_1$ we have

$$(11.1a.49) \qquad \sqrt{a}\,N^{\alpha} = \epsilon^{\alpha\lambda}\epsilon^{\mu\nu}M_{\lambda\nu|\mu} + \underset{\sim}{g}\ell^{\alpha} = m^{\alpha\mu}{}_{|\mu} + \underset{\sim}{g}\ell^{\alpha} .$$

The constitutive equations for an elastic, isotropic and homogeneous shell Günther obtained from the two-dimensional Hooke's law,

$$(11.1a.50) \qquad t^{\alpha\beta} = \frac{E}{1-\nu^2}\left[(1-\nu)g^{\alpha\beta}g^{\lambda\mu} + \nu g^{\alpha\lambda}g^{\beta\mu}\right]\gamma_{\lambda\mu} ,$$

where $\gamma_{\lambda\mu}$ is the strain tensor,

$$\gamma_{\lambda\mu} = \frac{1}{2}(g'_{\lambda\mu} - g_{\lambda\mu}) ,$$

and $g'_{\lambda\mu}$ is the deformed metric. If the points on the middle surface σ suffer a displacement $\underset{\sim}{u}$, from (11.1a.28) we have

$$(11.1a.51) \qquad \begin{aligned} g'_{\alpha\beta} &= a'_{\alpha\beta} - 2zb'_{\alpha\beta} + z^2c'_{\alpha\beta} , \\ c_{\alpha\beta} &\equiv \underset{\sim}{b}_{\alpha}\cdot\underset{\sim}{b}_{\beta} , \end{aligned}$$

where

$$
\begin{aligned}
a'_{\alpha\beta} &= a_{\alpha\beta} + 2\varepsilon_{\alpha\beta} , \\
\varepsilon_{\alpha\beta} &= 2\partial_{(\alpha} u_{\beta)} ,
\end{aligned}
\qquad (11.1a.52)
$$

and

$$
\boldsymbol{a}'_{\alpha} = \partial_{\alpha}(\boldsymbol{r} + \boldsymbol{u}) = \boldsymbol{a}_{\alpha} + \partial_{\alpha}\boldsymbol{u} . \qquad (11.1a.53)
$$

From (11.1a.53) and from $\boldsymbol{a}'_{\alpha}\cdot n' = 0$ we find for $\boldsymbol{n}'$ in the first approximation (for infinitesimal displacement gradients)

$$
\boldsymbol{n}' = \boldsymbol{n} - (\boldsymbol{n}\cdot\partial_{\alpha}\boldsymbol{u})\boldsymbol{a}^{\alpha} . \qquad (11.1a.54)
$$

From $(11.1a.15)_1$ we see that in the deformed configuration

$$
b'_{\alpha\beta} = -\partial_{\alpha}[\boldsymbol{n} - (\boldsymbol{n}\cdot\partial_{\lambda}\boldsymbol{u})\boldsymbol{a}^{\lambda}]\cdot(\boldsymbol{a}_{\beta} + \partial_{\beta}\boldsymbol{u})
$$

and when the products of the displacement gradients are neglected

$$
b'_{\alpha\beta} = b_{\alpha\beta} + (\partial_{\alpha}\partial_{\beta}\boldsymbol{u})\cdot\boldsymbol{n} \equiv b_{\alpha\beta} - \tilde{\varrho}_{\alpha\beta} \qquad (11.1a.55)
$$

where

$$
\tilde{\varrho}_{\alpha\beta} = -\boldsymbol{n}(\partial_{\alpha}\partial_{\beta}\boldsymbol{n}) \qquad (11.1a.56)
$$

represent the change of curvature. When the shell is in the initial configuration flat, i.e. when we consider a plate, the tensor $\tilde{\varrho}_{\alpha\beta}$ represents the curvature of the deformed plate.

Finally, from $(11.1a.51)_2$ we may write

$$c'_{\alpha\beta} = b'^{\ \lambda}_{\alpha} b'_{\beta\lambda} = a'^{\lambda\mu} b'_{\alpha\lambda} b'_{\beta\mu} \approx$$

(11.1a.57)

$$\approx c_{\alpha\beta} - (b^{\lambda}_{\alpha} \varrho_{\beta\lambda} + b^{\lambda}_{\beta} \varrho_{\alpha\lambda}) ,$$

where we have put

(11.1a.58) $$\varrho_{\alpha\beta} = \tilde{\varrho}_{\alpha\beta} + b^{\lambda}_{\alpha} \varepsilon_{\lambda\beta} ,$$

and $\varepsilon_{\lambda\beta}$ is the deformation tensor, $\varepsilon_{\lambda\beta} = \underset{\sim}{\varepsilon}_{\lambda} \cdot \underset{\sim}{a}_{\beta}$.

The strain tensor will be now

(11.1a.59) $$\gamma_{\alpha\gamma} = \frac{1}{2}(g'_{\alpha\beta} - g_{\alpha\beta}) = \varepsilon_{(\alpha\beta)} + z\tilde{\varrho}_{\alpha\beta} - \frac{1}{2} z^2 (b^{\lambda}_{\alpha} \varrho_{\beta\lambda} + b^{\lambda}_{\beta} \varrho_{\alpha\lambda}) .$$

From (11.1a.36) and (11.1a.50) we obtain for the reduced stress tensor the expression

(11.1a.60) $$\sigma^{\alpha\beta} = \frac{E}{1-\nu^2}\left[(1-\nu) g^{\alpha\lambda} (\underset{\sim}{g}^{\mu} \cdot \underset{\sim}{a}^{\beta}) + \nu g^{\lambda\mu} (\underset{\sim}{g}^{\alpha} \cdot \underset{\sim}{a}^{\beta})\right] \gamma_{\lambda\mu} .$$

Introducing $\gamma_{\alpha\beta}$ from (11.1a.59) and using (11.1a.46) we finally obtain the constitutive equations for the shell,

$$\sqrt{a} N^{\alpha\beta} = \frac{1-\nu^2}{E}\left\{(1-\nu) a^{\alpha\lambda} a^{\beta\mu}\left[\varepsilon_{(\lambda\mu)} + \frac{a^2}{12}\left(\frac{3}{2} b^{\nu}_{\lambda} \varrho_{\nu\mu} + \frac{1}{2} b^{\nu}_{\mu} \varrho_{\nu\lambda} - \right.\right.\right.$$

(11.1a.61)

$$\left.\left.\left. - 2H\varrho_{\lambda\mu}\right)\right] + \nu\left[a^{\alpha\beta}\varepsilon + \frac{a^2}{12}(a^{\alpha\beta} b^{\lambda\mu} \varrho_{\lambda\mu} + b^{\alpha\beta} \varrho - 2H a^{\alpha\beta} \varrho)\right]\right\} ,$$

$$m^{\alpha\beta} = \frac{Ea^3}{12(1-\nu^2)}\Big\{(1-\nu)a^{\alpha\lambda}a^{\beta\mu}\Big[\varrho_{\lambda\mu} + b^{\nu}_{\lambda}\varepsilon_{(\nu\mu)} + b^{\nu}_{\mu}\varepsilon_{(\nu\lambda)} - 2H\varepsilon_{(\lambda\mu)}\Big] + $$
$$+ \nu\big[a^{\alpha\beta}\varrho + a^{\alpha\beta}b^{\lambda\mu}\varepsilon_{(\lambda\mu)} + b^{\alpha\beta}\varepsilon - 2Ha^{\alpha\beta}\varepsilon\big]\Big\} . \qquad (11.1a.62)$$

Here we have put

$$H = \frac{1}{2}b^{\alpha}_{\alpha} , \quad \varepsilon = a^{\alpha\beta}\varepsilon_{\alpha\beta} , \quad \varrho = a^{\alpha\beta}\varrho_{\alpha\beta} . \qquad (11.1a.63)$$

11.1b Reissner's Theory

From the point of view of continuum mechanics, Reissner's approach to the theory of plates and shells [368–376] (cf. also Wan [479–471]) is based on the same kinematical model as Günther's theory (see equations (11.1a.1–5). Reissner's derivation of the shell equations differs from that of Günther in the approach to the problem of constitutive equations. Reissner developed an iteration procedure for deriving two-dimensional equations from an integro-differential formulation of the three-dimensional theory.

If we introduce into the fundamental equations of motion (7.37) and (7.41) the notation

$$\sqrt{g}\,\underset{\sim}{t}^{k} = \underset{\sim}{T}^{k} , \quad \sqrt{g}\,\underset{\sim}{\overset{*}{m}}^{k} = \underset{\sim}{\overset{*}{M}}^{k} ,$$
$$\varrho\sqrt{g}\,\underset{\sim}{f} = \underset{\sim}{p} , \quad \varrho\sqrt{g}\,\underset{\sim}{\overset{*}{l}} = \underset{\sim}{q} , \qquad (11.1b.1)$$

the equilibrium equations may be written in the vectorial form

(11.1b.2)
$$\partial_k \underset{\sim}{T}^k + \underset{\sim}{p} = 0 ,$$
$$\partial_k \overset{*}{\underset{\sim}{M}}{}^k + \underset{\sim}{g}_k \times \underset{\sim}{T}^k + \underset{\sim}{q} = 0 .$$

Two vectorial equations of equilibrium (11.1b.2), together with six compatibility conditions for Günther's deformation vectors (11.1a.1–2),

(11.1b.3)
$$\overset{(1)}{\underset{\sim}{I}}{}^k \equiv \epsilon^{klm} \partial_l \underset{\sim}{x}_m = 0 ,$$
$$\overset{(2)}{\underset{\sim}{I}}{}^k \equiv \epsilon^{klm} (\partial_l \underset{\sim}{\varepsilon}_m + \underset{\sim}{g}_l \times \underset{\sim}{x}_m) = 0 ,$$

represent the basic set of equations of Reissner's theory.

The faces of the shell are given by the equations

$$x^3 = z = \pm a(x^1, x^2) ,$$

where x^α, $\alpha = 1,2$ are coordinates on the middle surface σ of the shell, and $x^3 = z$ is orthogonal to σ. In the original papers Reissner chooses x^1, x^2 to be the lines of curvature of the middle surface. The face boundary conditions are

(11.1b.4)
$$z = \pm\frac{1}{2}a : \quad \underset{\sim}{T}^3 = 0, \quad \overset{*}{\underset{\sim}{M}}{}^3 = 0 .$$

Stress and couple resultants in the two-dimensional theory are assumed to be

$$\underset{\sim}{N}_\alpha = \int_{-\frac{a}{2}}^{\frac{a}{2}} \underset{\sim}{T}_\alpha dz , \quad \underset{\sim}{M}_\alpha = \int_{-\frac{a}{2}}^{\frac{a}{2}} (\overset{*}{\underset{\sim}{M}}_\alpha + z\underset{\sim}{n}\times\underset{\sim}{T}_\alpha) dz , \tag{11.1b.5}$$

where $\underset{\sim}{n}$ is again the unit normal vector to σ ,

$$\underset{\sim}{g}_3 = \underset{\sim}{n} = \underset{\sim}{a}_3 . \tag{11.1b.6}$$

The two-dimensional theory is obtained from the three dimensional theory by the systematic elimination of $x^3 = z$. We assume again that the position vector of any point of the shell is given by the relations of the form

$$\underset{\sim}{r}^* = \underset{\sim}{r}(x^1,x^2) + z\underset{\sim}{n}(x^1,x^2) , \tag{11.1b.7}$$

where $\underset{\sim}{r}$ is the position vector of points on σ . If $\underset{\sim}{a}_\alpha$ are base vectors on σ we have

$$\begin{gathered} \underset{\sim}{a}_\alpha = \partial_\alpha \underset{\sim}{r} \\ \underset{\sim}{g}_\alpha = \partial_\alpha \underset{\sim}{r}^* = \underset{\sim}{a}_\alpha + z\partial_\alpha \underset{\sim}{n} , \quad \underset{\sim}{g}_3 = \underset{\sim}{n} . \end{gathered} \tag{11.1b.8}$$

From the equilibrium equations (11.1b.2) we find

$$\frac{\partial \underset{\sim}{T}^3}{\partial z} = -\partial_\alpha \underset{\sim}{T}^\alpha - \underset{\sim}{p} \equiv -\underset{\sim}{R}(\underset{\sim}{T}^\alpha) - \underset{\sim}{p} , \tag{11.1b.9a}$$

(11.1b.9b) $$\frac{\partial \underset{\sim}{M}^3}{\partial z} = -\partial_\alpha \underset{\sim}{\overset{*}{M}}{}^\alpha - \underset{\sim}{g}_k \times \underset{\sim}{T}^k - \underset{\sim}{q} \equiv -\underset{\sim}{R}(\underset{\sim}{\overset{*}{M}}{}^\alpha) - \underset{\sim}{g}_k \times \underset{\sim}{T}^k - \underset{\sim}{q} .$$

Using the property of the sign-function $\mathrm{sgn}(x)$,

$$\frac{d}{dx}\mathrm{sgn}(x) = 2\delta(x) ,$$

where $\delta(x)$ is the delta-function, the integration of the two equations (11.1b.9) may be performed using the formulae *

(11.1b.10)
$$\underset{\sim}{T}^3 = \frac{1}{2}\int_{-\frac{a}{2}}^{\frac{a}{2}} \mathrm{sgn}(y-z)[\underset{\sim}{R}(\underset{\sim}{T}^3) + \underset{\sim}{p}]dy ,$$
$$\underset{\sim}{\overset{*}{M}}{}^3 = \frac{1}{2}\int_{-\frac{a}{2}}^{\frac{a}{2}} \mathrm{sgn}(y-z)[\underset{\sim}{R}(\underset{\sim}{\overset{*}{M}}{}^\alpha) + \underset{\sim}{q} + \underset{\sim}{g}_k \times \underset{\sim}{T}^k]dy .$$

Introducing now the values for $\underset{\sim}{T}^3$ and $\underset{\sim}{\overset{*}{M}}{}^3$ into the face boundary conditions $(11.1b.4)_1$, we obtain the relation

(11.1b.11) $$\underset{\sim}{T}^3\left(\pm\frac{a}{2}\right) = \frac{1}{2}\int_{-\frac{a}{2}}^{\frac{a}{2}} \mathrm{sgn}\left(y \pm \frac{a}{2}\right)[\underset{\sim}{R}(\underset{\sim}{T}^\alpha) + \underset{\sim}{p}]dy = 0 ,$$

---

* We use the following elementary properties of the integrals involving the delta-functions. a) $\delta(x-x_0) = 0$, $x \neq x_0$, b) $\int_{-\infty}^{\infty}\delta(x)dx = 1$, c) $\int_{-\infty}^{\infty} f(y)\delta(y-x)dy = f(x)$. Now, if $f'(x) = F(x)$, and if we write

$$f(x) = -\frac{1}{2}\int_{-\frac{a}{2}}^{\frac{a}{2}} \mathrm{sgn}(y-x)F(y)dy ,$$

by differentiation we obtain

$$f'(x) = \int_{-\frac{a}{2}}^{\frac{a}{2}} \delta(y-x)F(y)dy = F(x) .$$

which gives

$$\int_{-\frac{a}{2}}^{\frac{a}{2}} [\underset{\sim}{R}(\underset{\sim}{T}^{\alpha}) + \underset{\sim}{p}] dy = 0 . \qquad (11.1b.12)$$

Similarly, from $(11.1b.4)_2$ we obtain

$$\int_{-\frac{a}{2}}^{\frac{a}{2}} [\underset{\sim}{R}(\overset{*}{\underset{\sim}{M}}{}^{\alpha}) + \underset{\sim}{q} + \underset{\sim}{g}_k \times \underset{\sim}{T}^k] dy = 0 . \qquad (11.1b.13)$$

Remembering the relations (11.1b.5), we see that (11.1b.12) may be rewritten in the form

$$\partial_\alpha \underset{\sim}{N}^{\alpha} + \int_{-\frac{a}{2}}^{\frac{a}{2}} \underset{\sim}{p}\, dz = 0 \qquad (11.1b.14)$$

in which it represents the two-dimensional equilibrium equation for the resultant forces $\underset{\sim}{N}^{\alpha}$. Using $(11.1b.2)_1$ we obtain from (11.1b.13)

$$\int_{-\frac{a}{2}}^{\frac{a}{2}} [\partial_\alpha(\overset{*}{\underset{\sim}{M}}{}^{\alpha} + z\underset{\sim}{n}\times\underset{\sim}{T}^{\alpha}) + \underset{\sim}{a}_\alpha\times\underset{\sim}{T}^{\alpha} + z\underset{\sim}{n}\times\underset{\sim}{p} + \underset{\sim}{q} + \underset{\sim}{n}\times\partial_3\underset{\sim}{T}^3 + \underset{\sim}{n}\times\underset{\sim}{T}^3] dz = 0 .$$

However,

$$z\underset{\sim}{n}\times\partial_3\underset{\sim}{T}^3 = \partial_3(z\underset{\sim}{n}\times\underset{\sim}{T}^3) - \underset{\sim}{n}\times\underset{\sim}{T}^3$$

and in view of the face boundary conditions and $(11.1b.5)_2$ we finally have

$$(11.1b.15) \qquad \partial_\alpha \underset{\sim}{M}^\alpha + \int_{-\frac{a}{2}}^{\frac{a}{2}} \underset{\sim}{a}_\alpha \times \underset{\sim}{T}^\alpha dz + \int_{-\frac{a}{2}}^{\frac{a}{2}} (\underset{\sim}{q} + z\underset{\sim}{n} \times \underset{\sim}{p}) dz = 0 ,$$

which represents the two-dimensional equilibrium equation for resultant couple-stresses.

To obtain the two-dimensional deformation vectors we shall use the compatibility conditions (11.1b.3). To distinguish three-dimensional deformation vectors from the two-dimensional deformation vectors, we shall denote three- dimensional vectors by $\underset{\sim}{\varepsilon}_\alpha(z)$, $\underset{\sim}{\chi}_\alpha(z)$ and two-dimensional vectors by $\underset{\sim}{\varepsilon}_\alpha$, $\underset{\sim}{\chi}_\alpha$. From the first set of compatibility conditions we obtain two relations,

$$\partial_3 \underset{\sim}{\chi}_\alpha(z) = \partial_\alpha \underset{\sim}{\chi}_3(z) ,$$

and the integration gives

$$(11.1b.16) \qquad \underset{\sim}{\chi}_\alpha(z) = \underset{\sim}{\chi}_\alpha + \int_0^z \partial_\alpha \underset{\sim}{\chi}_3(z) dz .$$

From the second set of the compatibility conditions we also have two relations,

$$(11.1b.17) \qquad \partial_3 \underset{\sim}{\varepsilon}_\alpha(z) = \partial_\alpha \underset{\sim}{\varepsilon}_3(z) - \underset{\sim}{n} \times \underset{\sim}{\chi}_\alpha(z) + \underset{\sim}{g}_\alpha \times \underset{\sim}{\chi}_3(z)$$

which may be rewritten in the form

$$\partial_3 \underset{\sim}{\varepsilon}_\alpha(z) = \partial_\alpha \underset{\sim}{\varepsilon}_3(z) - \underset{\sim}{n}\times\underset{\sim}{\chi}_\alpha - \underset{\sim}{n}\times\int_0^z \partial_\alpha \underset{\sim}{\chi}_3(\eta)d\eta + {}$$

$$+ \underset{\sim}{a}_\alpha\times\underset{\sim}{\chi}_3(z) + z\partial_\alpha \underset{\sim}{n}\times\underset{\sim}{\chi}_3(z) .$$

Integrating this for $0 \leq \eta \leq z$ we obtain

$$\underset{\sim}{\varepsilon}_\alpha(z) = \underset{\sim}{\varepsilon}_\alpha - z\underset{\sim}{n}\times\underset{\sim}{\chi}_\alpha + \underset{\sim}{a}_\alpha\times\int_0^z \underset{\sim}{\chi}_3(\eta)d\eta + \int_0^z \eta\partial_\alpha \underset{\sim}{n}\times\underset{\sim}{\chi}_3(\eta)d\eta + {}$$

$$+ \int_0^z \partial_\alpha \underset{\sim}{\varepsilon}_3(\eta)d\eta - \int_0^z \left[\underset{\sim}{n}\times\int_0^y \partial_\alpha \underset{\sim}{\chi}_3(\eta)d\eta\right]dy .$$

Integration by parts gives

$$\int_0^z \left(\underset{\sim}{n}\times\int_0^y \partial_\alpha \underset{\sim}{\chi}_3(\eta)d\eta\right)dy = z\underset{\sim}{n}\times\int_0^z \partial_\alpha \underset{\sim}{\chi}(y)dy - \int_0^z y\underset{\sim}{n}\times\partial_\alpha \underset{\sim}{\chi}_3(y)dy ,$$

and we finally have

$$\underset{\sim}{\varepsilon}_\alpha(z) = \underset{\sim}{\varepsilon}_\alpha - z\underset{\sim}{n}\times\underset{\sim}{\chi}_\alpha + \underset{\sim}{a}_\alpha\times\int_0^z \underset{\sim}{\chi}_3(y)dy + {}$$
$$+ \int_0^z \left[\partial_\alpha \underset{\sim}{\varepsilon}_3(y) + y\partial_\alpha(\underset{\sim}{n}\times\underset{\sim}{\chi}_3(y)) - z\underset{\sim}{n}\times\partial_\alpha \underset{\sim}{\chi}_3(y)\right]dy . \qquad (11.1b.18)$$

Introducing (11.1b.16) and (11.1b.18) into the three-dimensional compatibility conditions which were not used in the derivation of (11.1b.16) and (11.1b.18), we obtain two two-dimensional compatibility conditions,

$$\partial_\alpha \underset{\sim}{x}_\beta - \partial_\beta \underset{\sim}{x}_\alpha = 0 \ ,$$

(11.1b.19)

$$\partial_\alpha \underset{\sim}{\varepsilon}_\beta - \partial_\beta \underset{\sim}{\varepsilon}_\alpha + \underset{\sim}{a}_\alpha \times \underset{\sim}{x}_\beta - \underset{\sim}{a}_\beta \times \underset{\sim}{x}_\alpha = 0 \ .$$

To obtain the components of the deformation tensors we shall consider the scalar products of the vectors $\underset{\sim}{\varepsilon}_\alpha(z)$ and $\underset{\sim}{x}_\alpha(z)$ in the relations (11.1b.16) and (11.1b.18) with the base vectors $\underset{\sim}{a}_\alpha$ and $\underset{\sim}{a}_3 = \underset{\sim}{n}$ (Reissner does not use the base vectors, but the unit tangent vectors to the lines of curvature on σ),

$$\underset{\sim}{x}_{\alpha i}(z) = x_{\alpha i} + \int_0^z T_{\alpha i}[\underset{\sim}{x}_{3i}(y)]dy \ ,$$

$$(\alpha = 1,2 \ ; \ \ i = 1,2,3)$$

(11.1b.20)

$$\varepsilon_{\alpha i}(z) = \varepsilon_{\alpha i} + z\epsilon_{i\gamma 3} x_\alpha^{\cdot\gamma} - \epsilon_{\alpha i 3}\int_0^z x_{33}(y)dy \ +$$

$$+ \int_0^z S_{\alpha i}[\varepsilon_{3i}(y), x_{3i}(y)]dy \ ,$$

$$\epsilon_{\alpha i 3} = (\underset{\sim}{a}_\alpha \times \underset{\sim}{a}_i)\underset{\sim}{n} \ .$$

Components of the stress vector $\underset{\sim}{T}^{3}$ may be obtained applying the same procedure to $(11.1b.10)_1$,

$$T^{i3}(z) = \underset{\sim}{T}^{3}\cdot\underset{\sim}{a}^{i} = \frac{1}{2}\int_{-\frac{a}{2}}^{\frac{a}{2}} \operatorname{sgn}(y-z)[R^{i}(\underset{\sim}{T}^{\alpha})+p^{i}]dy \ , \qquad (11.1b.21)$$

where

$$R^{i}(\underset{\sim}{T}^{\alpha}) = \underset{\sim}{R}(\underset{\sim}{T}^{\alpha})\cdot\underset{\sim}{a}^{i} = R^{i}[T^{\alpha i}(y)] \ ,$$

and for the components of the couples $\overset{*}{M}{}^{3}$ we obtain

$$\overset{*}{M}{}^{i3}(z) = \frac{1}{2}\int_{-\frac{a}{2}}^{\frac{a}{2}} \operatorname{sgn}(y-z)[R^{i}(\overset{*}{\underset{\sim}{M}}{}^{\alpha}) + q^{i} + (\underset{\sim}{a}^{i}\times\underset{\sim}{g}_{k})\cdot\underset{\sim}{T}^{k}]dy \ . \qquad (11.1b.22)$$

In (11.1b.20) we have the expressions for twelve components of the deformation $\varkappa_{\alpha i}(z)$, $\varepsilon_{\alpha i}(z)$, expressed in terms of the components of $\varepsilon_{3i}(z)$, $\varkappa_{3i}(z)$, through certain integral relations.

For three-dimensional elastic bodies the linear constitutive relations are of the form

$$\begin{aligned} \varepsilon_{ij}(z) &= \overset{(1)}{C}_{ijkl}T^{kl}(z) + \overset{(2)}{C}_{ijkl}\overset{*}{M}{}^{kl}(z) \ , \\ \varkappa_{ij}(z) &= \overset{(1)}{D}_{ijkl}T^{kl}(z) + \overset{(2)}{D}_{ijkl}\overset{*}{M}{}^{kl}(z) \ . \end{aligned} \qquad (11.1b.23)$$

Introduction of the strain and stress components from (11.1b.20–22) into these constitutive equations yields a system of eighteen integral equations for the determination of

$T^{i\alpha}$, $\overset{*}{M}{}^{i\alpha}$; ε_{3i} and $\varkappa_{3i}$ as functions of z . Together with the six two-dimensional equilibrium equations (11.1b.14, 15) and with the six two-dimensional compatibility-conditions (11.1b.16, 17) we thus have a system of thirty integrodifferential equations for thirty quantities, among which twelve quantities $\varepsilon_{\alpha i}$ and $\varkappa_{\alpha i}$ do not depend on z .

As an illustration of these integro-differential equations, we shall write only three of them, and for transversally isotropic material for which Reissner assumes the linear stress-strain relations

$$'\varepsilon_{\alpha\alpha} = \frac{1}{E}[T_{\alpha\alpha} - \nu(I_T - T_{\alpha\alpha})] - \frac{\nu_z}{E_z}T_{33} ,$$

(11.1b.24) $$\varepsilon_{\alpha\beta} = \frac{1+\nu}{E}T_{\alpha\beta} , \quad (\alpha \neq \beta)$$

$$\varepsilon_{\alpha 3} = \frac{1}{G}T_{\alpha 3} , \quad \varepsilon_{3\alpha} = \frac{1}{G}T_{3\alpha} , \quad \varepsilon_{33} = \frac{1}{E_z}(T_{33} - \nu I_t) , \quad I_T \equiv T_{11} + T_{22}$$

$$\varkappa_{\alpha\beta} = \frac{1}{h^2 F}\overset{*}{M}_{\alpha\beta} , \quad \varkappa_{\alpha 3} = \frac{1}{h^2 H}\overset{*}{M}_{\alpha 3} , \quad \varkappa_{3\alpha} = \frac{1}{h^2 H_z}\overset{*}{M}_{3\alpha} ,$$

(11.1b.25)

$$\varkappa_{33} = \frac{1}{h^2 F_z}\overset{*}{M}_{33} .$$

E.g. we have

$$\varepsilon_{11} + z x_1^{\cdot 2} + \int_0^z S_{11} dy = \frac{1}{E} T_{11} - \frac{\nu}{E} T_{22} - \frac{\nu_z}{2E_z} \int_0^z \mathrm{sgn}(y-z)[k^3 + p^3] dy \; ,$$

$$x_{11} + \int_0^z T_{11} dy = \frac{1}{h^2 F} \overset{*}{M}_{11}, \ldots \; ; \text{ etc.}$$

Further elaboration of the iteration and approximation methods to be applied to the integro-differential equations of this shell theory is beyond the scope of this exposition. We shall only notice here that in the theory of shallow shells (Reissner and Wan [376] , Reissner [371] , Wan [479]) the shell is considered as a surface, but the kinematical model is the same as in the general theory. The theory of shallow shells is completely two-dimensional and does not involve the integro-differential equations of the general theory.

11.2 Theories with Deformable Directors

The theories of plates, shells and rods with deformable directors are based on the assumption that the three--dimensional material is an ordinary material in the classical sense, and the appearance of the directors in the theory is a result of the reduction of the three-dimensional theory to a one or two-dimensional theory. We have already reviewed some of the basic ideas and relations in the sections 5.5.1, 5.5.2, 7.2 and 7.3 but in those sections we have not considered the constitutive equations.

It seems to me that the most general approach is offered by the theory of Green, Naghdi and Laws [157, 169]. In this section we shall give only an outline of their treatment of the subject.

We consider the energy balance law in the form (8.1.1), assuming that there are no volume couples acting on the points of the body and that the material is non-polar. For the kinetic energy we use (5.5.1.23),

$$\text{(11.2.1)} \quad \frac{d}{dt}\int_{v} \varrho^*\left(\frac{1}{2}\underset{\sim}{v}^*\cdot\underset{\sim}{v}^* + \varepsilon^*\right)dv = \oint_{s}(\underset{\sim}{t}\cdot\underset{\sim}{v}^* + q^*)ds + \int_{v}\varrho^*(\underset{\sim}{f}\cdot\underset{\sim}{v}^* + h^*)dv .$$

For the part v of the shell we choose a cylinder defined by a closed contour C on the middle surface, and by the surfaces $X = \alpha$ and $X = \beta$. The element of an area of the surface $X = \text{const}$ is

$$\text{(11.2.2)} \quad |(\underset{\sim}{g}_1\times\underset{\sim}{g}_2|dX^1dX^2 = |g_{11}g_{22} - g_{12}g_{21}|dX^1dX^2 = \sqrt{gg^{33}}\,dX^1dX^2 .$$

For the surface integral on the right-hand side of (11.2.1) we may write now

$$\oint_{s} H ds = \int_{A} H dC dX + \left[H\sqrt{gg^{33}}\right]_{X=\alpha} dX^1dX^2 + \left[H\sqrt{gg^{33}}\right]_{X=\beta} dX^1dX^2 ,$$

where A is the cylindrical surface determined by the contour C and dC is the arc element of C. Thus we may write

$$\oint_S \underset{\sim}{t}\, ds = \int_A \underset{\sim}{t}\, dA + \left(\left[\underset{\sim}{t}\sqrt{g g^{33}}\right]_{X=\alpha} + \left[\underset{\sim}{t}\sqrt{g g^{33}}\right]_{X=\beta}\right) dX^1 dX^2 . \qquad (11.2.3)$$

For $X = \alpha$ the unit normal vector is $\underset{\sim}{n} = \underset{\sim}{g}^3/\sqrt{g^{33}}$ and therefore

$$\underset{\sim}{t} = \underset{\sim}{t}^i(\underset{\sim}{g}^3\cdot\underset{\sim}{g}_i)/\sqrt{g^{33}} = \underset{\sim}{t}^3/\sqrt{g^{33}} .$$

For $X = \beta$ we have $\underset{\sim}{t} = -\underset{\sim}{t}^3/\sqrt{g^{33}}$ and

$$\oint_S \underset{\sim}{t}\, ds = \int_A \underset{\sim}{t}\, dC dX + \left[\sqrt{g}\,\underset{\sim}{t}^3\right]_\alpha^\beta dX^1 dX^2 . \qquad (11.2.4)$$

Similarly we obtain

$$\oint_S q^* dS = \int_A q^* dC dX + \left(\left[q^*\sqrt{g g^{33}}\right]_{X=\alpha} + \left[q^*\sqrt{g g^{33}}\right]_{X=\beta}\right) dX^1 dX^2 . \qquad (11.2.5)$$

If we put in (11.2.2) $H = H^\alpha n_\alpha$, where $\underset{\sim}{n}$ is the unit outward normal to the surface A, we have to use the following relation

$$n_\alpha dA = \underset{\sim}{g}_\alpha\cdot\underset{\sim}{n}\, dA = \underset{\sim}{g}_\alpha\cdot(\underset{\sim}{\tau}\times\underset{\sim}{g}_3) dC dX ,$$

where $\underset{\sim}{\tau}$ is the unit tangent vector to the curve C, and $\underset{\sim}{\tau}\, dC = dx^\beta \underset{\sim}{g}_\beta$. Further

$$n_\alpha dA = (\underset{\sim}{g}_3\times\underset{\sim}{g}_\alpha)\cdot\underset{\sim}{g}_\beta dX^\beta = \sqrt{g}(e_{3\alpha 1} dX^1 + e_{3\alpha 2} dX^2) . \qquad (11.2.6)$$

Thus

$$\int_A H dA = \oint_C \int_\alpha^\beta (\sqrt{g} H^1 dX^2 - \sqrt{g} H^2 dX^1) dX .$$

If we put $\int_\alpha^\beta \sqrt{g} H^\alpha dX = \tilde{H}^\alpha$, and in the analogy to (11.2.6), at the intersection of the surfaces A and $X = 0$ denote by $\underset{\sim}{\nu}$ the normal to C, we have

$$\nu_1 dC = \sqrt{a}\, dX^2, \qquad \nu_2 dC = -\sqrt{a}\, dX^1. \tag{11.2.7}$$

Using this we finally obtain

$$\int_A H dA = \oint_C \tilde{H}^\alpha \nu_\alpha dC = \oint_C \tilde{H} dC. \tag{11.2.8}$$

We introduce now the notation

$$\int_\alpha^\beta \varrho^* \underset{\sim}{f}^* \sqrt{g}\, dX + \left[\sqrt{g}\, \underset{\sim}{t}_3\right]_\alpha^\beta = \varrho \underset{\sim}{F} \sqrt{a}, \tag{11.2.9}$$

$$\int_\alpha^\beta \varrho^* h^* \sqrt{g}\, dX + \left[\sqrt{g g^{33}}\, q^*\right]_{X=\alpha} + \left[\sqrt{g g^{33}}\, q^*\right]_{X=\beta} = \varrho h \sqrt{a}, \tag{11.2.10}$$

$$\int_\alpha^\beta \underset{\sim}{t}_\alpha \sqrt{g}\, dX = \underset{\sim}{N}_\alpha \sqrt{a}, \qquad (\underset{\sim}{N} = \underset{\sim}{N}^\alpha \nu_\alpha) \tag{11.2.11}$$

$$\int_\alpha^\beta X^N \sqrt{g}\, \underset{\sim}{t}_\alpha dX = \underset{\sim}{M}^N_\alpha \sqrt{a}, \qquad (\underset{\sim}{M}^N = \underset{\sim}{M}^{N\alpha} \nu_\alpha) \tag{11.2.12}$$

$$\int_\alpha^\beta q^{*\alpha} \sqrt{g}\, dX = q^\alpha \sqrt{a}, \qquad (q = q^\alpha \nu_\alpha) \tag{11.2.13}$$

$$\int_\alpha^\beta \varrho^* \varepsilon^* \sqrt{g}\, dX = \varrho \varepsilon \sqrt{a}, \tag{11.2.14}$$

$$\int_{\alpha}^{\beta} \varrho^* \underset{\sim}{f}^* X^N \sqrt{g}\, dX + \left[\underset{\sim}{t} X^N \sqrt{g g^{33}}\right]_{X=\alpha} + \left[\underset{\sim}{t} X^N \sqrt{g g^{33}}\right]_{X=\beta} = \varrho \underset{\sim}{L}^N \sqrt{a} \,. \quad (11.2.14a)$$

Using now the formulae (11.2.2–13) we obtain from (11.2.1) the following expression for the energy balance law,

$$\frac{d}{dt}\int_{\sigma} \varrho\left(\varepsilon + \frac{1}{2}\underset{\sim}{v}\cdot\underset{\sim}{v} + \sum_{N=2}^{\infty} k^N \underset{\sim}{v}\cdot\dot{\underset{\sim}{d}}_{(N)} + \frac{1}{2}\sum_{M,N=1}^{\infty} k^{MN} \dot{d}_{(M)}\cdot\dot{d}\right) d\sigma = \quad (11.2.15)$$

$$= \int_{\sigma} \varrho\left(h + \underset{\sim}{F}\cdot\underset{\sim}{v} + \sum_{N=1}^{\infty} \underset{\sim}{L}^N \dot{\underset{\sim}{d}}_{(N)}\right) d\sigma + \oint_{C}\left(\underset{\sim}{N}\cdot\underset{\sim}{v} + \sum_{N=1} \underset{\sim}{M}^N \dot{\underset{\sim}{d}}_{(N)} - q\right) dC \,.$$

This expression is completely two-dimensional.

From the invariance of the energy balance law under superposed rigid body motions we obtain, using (11.2.15), the equations of motion and the simplified energy equation. Following the procedure of the section 8.1 we obtain the following equations,

$$\dot{\varrho} + \varrho(v^{\alpha}{}_{|\alpha} - v^3 b^{\alpha}_{\alpha}) = 0 \,, \quad (11.2.16)$$

$$\varrho \dot{\underset{\sim}{v}} = \underset{\sim}{N}^{\alpha}{}_{|\alpha} + \varrho \underset{\sim}{F} - \sum_{N=2}^{\infty} k^N \ddot{\underset{\sim}{d}}_{(N)} \,, \quad (11.2.17)$$

$$\underset{\sim}{N}^{\alpha} \times \underset{\sim}{a}_{\alpha} + \sum_{N=1}^{\infty}\left(\underset{\sim}{m}^N \times \underset{\sim}{d}_{(N)} + \underset{\sim}{M}^{N\alpha} \times \underset{\sim}{d}_{(N)|\alpha}\right) = 0 \,, \quad (11.2.18)$$

where

(11.2.19) $$\underset{\sim}{m}^N\sqrt{a} = N\int_\alpha^\beta X^{N-1}\underset{\sim}{t}_3\sqrt{g}\,dX\ .$$

Using the equations of motion (11.2.16–19) the energy equation may be reduced to the simpler form,

(11.2.10) $$\varrho\dot{\varepsilon} + N^{|\beta\alpha}\dot{a}_{\beta\alpha} + \sum_{N=1}^{\infty}\underset{\sim}{m}^{Ni}\dot{\underset{\sim}{d}}_{(N)i} + \sum_{N=1}^{\infty}M^{Ni\alpha}\dot{\lambda}_{Ni\alpha} + \varrho h + q^{\alpha}_{|\alpha} = 0\ ,$$

where

(11.2.21)
$$N^{|\beta\alpha} = N^{\beta\alpha} - \sum_{N=1}^{\infty}(m^{N\alpha}d^{\beta}_{(N)} + M^{N\alpha\gamma}\lambda_{N\cdot\gamma}^{\cdot\beta})\ ,$$

$$\lambda_{N\beta\alpha} = d_{(N)\beta|\alpha} - b_{\beta\alpha}d_{(N)3}\ ,$$

$$\lambda_{N3\alpha} = \partial_\alpha d_{(N)3} + b^{\beta}_{\alpha}d_{(N)\beta}\ ,$$

$$\dot{a}_{\alpha\beta} = 2\dot{e}_{\alpha\beta}\ .$$

If we introduce the free energy function $\Psi = \varepsilon - \eta\theta$, and if we assume

$$\Psi = \Psi(\theta\,,\, e_{\alpha\beta}\,,\, \lambda_{Ni\alpha}\,,\, d_{(N)i})$$

following in principle the procedure of the section 10 we obtain the constitutive equations of the two dimensional shell theory,

$$\eta = -\frac{\partial\Psi}{\partial\theta},$$

$$N^{|\alpha\beta} = \varrho\frac{\partial\Psi}{\partial e_{\alpha\beta}}, \quad m^{Ni} = \varrho\frac{\partial\Psi}{\partial d_{(N)i}}, \quad M^{Ni\alpha} = \varrho\frac{\partial\Psi}{\partial\lambda_{Ni\alpha}}. \tag{11.2.22}$$

Without entering deeper into the details of this shell theory we shall only mention that the constitutive equations resemble very much the constitutive equations of the theory of micromorphic media (10.4.6). The appearance of the directors and of the director gradients here is natural consequence of the reduction of the three-dimensional theory to two dimensions. For details of the theory we refer the reader to the original papers [157, 169]. In the last of these papers Green and Naghdi have developed a general, non-isothermal theory.

11.3 Rods

The general theory of rods of Green, Naghdi and Laws [157, 169] is in essence based on the same ideas as were the ideas in the just outlined theory of shells. The fundamental quantities are already derived in the section 5.5.2, and for the first approximation (in which a rod is considered as a line with two directors) in the section 7.3.

The one-dimensional form of the energy equation

(11.2.1) for a part $\Phi_1 \leq X \leq \Phi_2$ of a rod, where $X = X^3$ is the parameter varying along the middle line, is (we are using the notation of the section 5.5.2)

$$\frac{d}{dt}\int_{\Phi_1}^{\Phi_2} \varrho\Big(\varepsilon + \frac{1}{2}\underset{\sim}{v}\cdot\underset{\sim}{v} + \sum_{N=2}^{\infty} k^{\alpha_1\ldots\alpha_N}\underset{\sim}{v}\cdot\dot{\underset{\sim}{d}}_{\alpha_1\ldots\alpha_N} +$$

$$+ \sum_{M,N=1}^{\infty} k^{\alpha_1\ldots\alpha_N\beta_1\ldots\beta_M}\dot{d}_{\alpha_1\ldots\alpha_N}\cdot\dot{d}_{\beta_1\ldots\beta_M}\Big)\sqrt{a_{33}}\,dX =$$

(11.3.1)

$$= \int_{\Phi_1}^{\Phi_2} \varrho\Big(\underset{\sim}{f}\cdot\underset{\sim}{v} + \sum_{N=1}^{\infty} \underset{\sim}{\ell}^{\alpha_1\ldots\alpha_N}\cdot\dot{\underset{\sim}{d}}_{\alpha_1\ldots\alpha_N} + h\Big)\sqrt{a_{33}}\,dX +$$

$$+ \Big[\underset{\sim}{n}\cdot\underset{\sim}{v} + \sum_{N=1}^{\infty} \underset{\sim}{p}^{\alpha_1\ldots\alpha_N}\cdot\dot{\underset{\sim}{d}}_{\alpha_1\ldots\alpha_N} + q\Big]_{\Phi_1}^{\Phi_2} .$$

Here

$$\iint \varrho^*\sqrt{g}\,dX^1dX^2 = \iint k\,dX^1dX^2 = \varrho\sqrt{a_{33}} ,$$

$$\iint kX^{\alpha_1}\ldots X^{\alpha_N}dX^1dX^2 = \varrho k^{\alpha_1\ldots\alpha_N}\sqrt{a_{33}} ,$$

$$\iint kh^*dX^1dX^2 + \oint q^*(n_1dX^2 - n_2dX^1)\sqrt{g} = \varrho h\sqrt{a_{33}} ,$$

$$\iint k\underset{\sim}{f}^{*}dX^1dX^2+\oint\sqrt{g}(\underset{\sim}{t}_1dx^2-\underset{\sim}{t}_2dx^1)=\varrho\underset{\sim}{f}\sqrt{a_{33}}\ ,$$

$$\iint k\underset{\sim}{f}^{*}X^{\alpha_1}\ldots X^{\alpha_N}dX^1dX^2+\oint\sqrt{g}X^{\alpha_1}\ldots X^{\alpha_N}(\underset{\sim}{t}_1dX^2-\underset{\sim}{t}_2dX^1)=\varrho\underset{\sim}{\ell}^{\alpha_1\ldots\alpha_N}\sqrt{a_{33}}\ .$$

$$\iint\underset{\sim}{t}_3\sqrt{g}dX^1dX^2=\underset{\sim}{n}\ ;\quad \iint X^{\alpha_1}\ldots X^{\alpha_N}\underset{\sim}{t}_3\sqrt{g}dX^1dX^2=\underset{\sim}{p}^{\alpha_1\ldots\alpha_N}\ .$$

The double integrals are over any cross-section $X = \text{const}$ of the rod, bounded by the curve (5.5.2.11), and the line-integral is along the curve defined by (5.5.2.11) and $X = \text{const}$.

From (11.3.1) the rod equations may be derived following the procedure analogous to that applied in the preceding section to the shells and we refer the interested reader to the original papers by Green, Naghdi and Laws.

11.4 Laminated Composite Materials

Laminated composites represent because of their practical engineering interest an important field of applications of the theory of materials with directors. In parallel layers each layer might be considered as a uni-directorial micro-element. This point of view was adopted by Hermann and

Achenbach, who developed a general dynamic theory of laminated composites. Details of their theory are beyond the scope of this course of lectures and we refer the readers to the original papers [1, 2, 3, 203] where further references may be found.

12. Polar Fluids

In comparison with the theory of elasticity, the theory of polar fluids is considerably less developed, although there are certain effects predicted by the theory which might be experimentally observed.

The flow of a fluid, if it is not an "ideal" fluid, is a dissipative process and the constitutive equations cannot be directly derived from the laws of thermodynamics, as was the case with the theory of elasticity.

The equations of motion,

$$\rho \ddot{x}^{i} = t^{ij}{}_{,j} + \rho f^{i} , \tag{12.1}$$

$$\rho i^{\lambda\mu} \ddot{d}^{i}_{(\mu)} = h^{(\lambda)ij}{}_{,j} + \rho k^{(\lambda)i} , \tag{12.2}$$

$$\rho \dot{\sigma}^{ij} = t^{[ij]} + \overset{*}{m}{}^{ijk}{}_{,k} + \rho \overset{*}{l}{}^{ij} \tag{12.3}$$

have a general validity, independently of the consistency of the material. These equations do not impose any restrictions on the constitutive equations. But the laws of thermodynamics,

$$\varrho\dot{\varepsilon} = w + \varrho h + q^{k}{}_{,k} , \quad (12.4)$$

$$\varrho\theta\dot{\eta} - \varrho h - q^{k}{}_{,k} + \frac{1}{\theta}\theta_{,k}q^{k} \geqslant 0 , \quad (12.5)$$

impose certain restrictions, since the constitutive equations cannot violate them.

The general scheme to be followed in the formulation of the theory of polar fluids might be considered as the following one. First, select a mechanical model and the appropriate kinematical variables, and then postulate constitutive equations and see that they are in agreement with the laws of thermodynamics.

There are today two main concepts of polar fluids, besides the theory of liquid crystals and anisotropic fluids which might be considered as a special case of a general theory of "generalized Cosserat fluids" which does not exist yet. Both theories predict certain effects which are expected to give an experimental evidence of the influence of the non-symmetric stress upon the distribution of velocities.

In the following subsections we shall give a brief review of these theories.

12.1 Micropolar Fluids

The basis of the theory of micropolar fluids represents the general concept of a micromorphic medium, which was introduced into fluid mechanics first by Eringen [124] parallel with the development of the theory of micromorphic elasticity, and later further developed in a series of papers * Eringen [123, 125, 125, 133, 135] , Eringen and Ingram [137], Allen, DeSilva and Kline [10,11], Allen and Kline [12], Ariman [15], Ariman and Cakmak [16, 17, 18] , Condif and Dahler [69], Kirwan and Newman [233, 234] , Kline [235] , Kline and Allen [236, 237, 238] , Liu [272], Rao et al. [367] etc.).

Quantities which characterize the state of stress in a micromorphic medium (cf. section 10.4) are the stress tensor t^{ij} , the micro-stress average tensor s^{ij} and the first stress moment tensor λ^{ijk} . The rates, according to our notation, are: velocity gradients $v_{i,j}$, gyration ω_{ij} , and the gyration gradients $\omega_{ij,k}$. If the phenomena including the heat conduction are excluded from the considerations, there are nineteen unknowns which have to be determined through the equations of motion:

$$\varrho(\underset{\sim}{x}, t), \quad I^{\ell m}(\underset{\sim}{x}, t), \quad v^{k}(\underset{\sim}{x}, t), \quad \omega_{k\ell}(\underset{\sim}{x}, t).$$

---

* A similar theory was independently developed by Aero, Bul'gin and Kuvshinskii [4].

The principle of objectivity requires that the constitutive variables are objective tensors. Such tensors are the rate of deformation d_{ij} and the micro-deformation rate tensors $\underset{\sim}{b}$ and $\underset{\sim}{a}$,

$$d_{ij} = v_{(i,j)} ,$$

$$b_{ij} = \omega_{ij} + v_{i,j} , \qquad (12.1.1)$$

$$a_{ijk} = \omega_{ij,k} .$$

According to Eringen [124] , a fluid is a micro-fluid if its constitutive equations are of the form

$$\underset{\sim}{t} = \underset{\sim}{f}(v_{i,j} , \omega_{ij} , \omega_{ij,k}) ,$$

$$\underset{\sim}{s} = \underset{\sim}{g}(v_{i,j} , \omega_{ij} , \omega_{ij,k}) , \qquad (12.1.2)$$

$$\underset{\sim}{\lambda} = \underset{\sim}{h}(v_{i,j} , \omega_{ij} , \omega_{ij,k}) ,$$

subject to the spatial and material objectivity and

$$\underset{\sim}{t} = \underset{\sim}{s} = -\pi \underset{\sim}{1} , \quad \underset{\sim}{\lambda} = 0 \qquad (12.1.3)$$

when $d_{ij} = v_{(i,j)} = 0$ and $b_{ij} = \omega_{ij} + v_{i,j} = 0$.

Another assumption which is made in the theory of fluids with micro-structure is that the fluid possesses an inter nal energy ε which depends solely on the entropy η, specific volume $1/\varrho$ and on the micro-inertia I^{km},

$$\varepsilon = \varepsilon(\eta, \varrho^{-1}, I^{km}). \tag{12.1.4}$$

With this we may define the following quantities, thermodynamic temperature θ, thermodynamic pressure π and thermodynamic micro-pressure π_{ij},

$$\theta = \left.\frac{\partial\varepsilon}{\partial\eta}\right|_{\varrho,\underset{\sim}{I}=\text{const}} \qquad \pi = \left.\frac{\partial\varepsilon}{\partial\varrho^{-1}}\right|_{\eta,\underset{\sim}{I}=\text{const}}$$

$$\pi_{ij} = \left.\frac{\partial\varepsilon}{\partial I^{ij}}\right|_{\eta,\varrho=\text{const}} . \tag{12.1.5}$$

For the constitutive equations we shall write now

$$\underset{\sim}{t} = \underset{\sim}{f}(\underset{\sim}{d}, \underset{\sim}{b}, \underset{\sim}{a}), \quad \underset{\sim}{s} = \underset{\sim}{g}(\underset{\sim}{d}, \underset{\sim}{b}, \underset{\sim}{a}), \quad \underset{\sim}{\lambda} = \underset{\sim}{h}(\underset{\sim}{d}, \underset{\sim}{b}, \underset{\sim}{a}) \tag{12.1.6}$$

where $\underset{\sim}{t}$ and $\underset{\sim}{s}$ are second-order tensors, and $\underset{\sim}{\lambda}$ is a third-order tensor. The principle of objectivity requires that

$$\underset{\sim}{f}(\underset{\sim}{Q}\underset{\sim}{d}\underset{\sim}{Q}^T, \underset{\sim}{Q}\underset{\sim}{b}\underset{\sim}{Q}^T, \underset{\sim}{Q}\underset{\sim}{a}\underset{\sim}{Q}^T\underset{\sim}{Q}^T) = \underset{\sim}{Q}\underset{\sim}{f}\underset{\sim}{Q}^T ,$$

and similarly for $\underset{\sim}{g}$, and for $\underset{\sim}{h}$,

$$\underset{\sim}{h}(\underset{\sim}{Q}\underset{\sim}{d}\underset{\sim}{Q}^T, \underset{\sim}{Q}\underset{\sim}{b}\underset{\sim}{Q}^T, \underset{\sim}{Q}\underset{\sim}{a}\underset{\sim}{Q}^T\underset{\sim}{Q}^T) = \underset{\sim}{Q}\underset{\sim}{h}\underset{\sim}{Q}^T ,$$

where $\underset{\sim}{Q}$ is an arbitray orthogonal matrix. If we select $\underset{\sim}{Q} = -\underset{\sim}{1}$ we will obtain

$$
\begin{aligned}
\underset{\sim}{f}(\underset{\sim}{d},\underset{\sim}{b},-\underset{\sim}{a}) &= \underset{\sim}{f}(\underset{\sim}{d},\underset{\sim}{b},\underset{\sim}{a}) \\
\underset{\sim}{g}(\underset{\sim}{d},\underset{\sim}{b},-\underset{\sim}{a}) &= \underset{\sim}{g}(\underset{\sim}{d},\underset{\sim}{b},\underset{\sim}{a}) \qquad (12.1.7) \\
\underset{\sim}{h}(\underset{\sim}{d},\underset{\sim}{b},-\underset{\sim}{a}) &= -\underset{\sim}{h}(\underset{\sim}{d},\underset{\sim}{b},\underset{\sim}{a})
\end{aligned}
$$

and it follows that $\underset{\sim}{f}$ and $\underset{\sim}{g}$ have to be even functions, and $\underset{\sim}{h}$ an odd function in $\underset{\sim}{a}$.

The general constitutive equations that were considered by Eringen [124] were

$$
\begin{aligned}
t^{k\ell} &= \underset{0}{f}^{k\ell}(\underset{\sim}{d},\underset{\sim}{b}-\underset{\sim}{d},\underset{\sim}{b}^T-\underset{\sim}{d}) + O(\underset{\sim}{a}^2), \\
s^{k\ell} &= \underset{0}{g}^{k\ell}(\underset{\sim}{d},\underset{\sim}{b}-\underset{\sim}{d},\underset{\sim}{b}^T-\underset{\sim}{d}) + O(\underset{\sim}{a}^2), \qquad (12.1.8) \\
\lambda^{k\ell m} &= \underset{0}{h}^{k\ell m}(\underset{\sim}{d},\underset{\sim}{b}-\underset{\sim}{d},\underset{\sim}{b}^T-\underset{\sim}{d}) + O(\underset{\sim}{a}^3)
\end{aligned}
$$

where $\underset{\sim}{b}-\underset{\sim}{d}$ and $\underset{\sim}{b}^T-\underset{\sim}{d}$ are introduced instead of $\underset{\sim}{b}$ for later convenience.

According to (12.1.3) we may add that for the vanishing $\underset{\sim}{d}$ and $\underset{\sim}{b}$ the right-hand sides of (12.1.8) have to satisfy the following conditions,

$$f^{k}_{.\ell}(0,0,0) = -\pi\delta^k_\ell , \quad g^{k}_{.\ell}(0,0,0) = -\pi\delta^k_\ell ,$$

(12.1.9)

$$\lambda^{k}_{.\ell m}(0,0,0) = 0 .$$

Taking all this into account, in the linear approximation the constitutive equations for micro-fluids are

$$\underset{\sim}{t} = [-\pi + \lambda_v \operatorname{tr} \underset{\sim}{d} + \lambda_0 \operatorname{tr}(\underset{\sim}{b} - \underset{\sim}{d})]\underset{\sim}{1} + 2\mu_v \underset{\sim}{d} + 2\mu_0(\underset{\sim}{b} - \underset{\sim}{d}) + 2\mu_1(\underset{\sim}{b}^T - \underset{\sim}{d}) ,$$

(12.1.10)

$$\underset{\sim}{s} = [-\pi + \eta_v \operatorname{tr} \underset{\sim}{d} + \eta_0 \operatorname{tr}(\underset{\sim}{b} - \underset{\sim}{d})]\underset{\sim}{1} + 2\zeta_v \underset{\sim}{d} + \zeta_1(\underset{\sim}{b} + \underset{\sim}{b}^T - 2\underset{\sim}{d}) .$$

$$\lambda_{k\ell m} = (\gamma_1 a_{mrr} + \gamma_2 a_{rmr} + \gamma_3 a_{rrm})\delta_{k\ell} +$$

$$+(\gamma_4 a_{\ell rr} + \gamma_5 a_{r\ell r} + \gamma_6 a_{rr\ell})\delta_{km} +$$

(12.1.11)

$$+(\gamma_7 a_{krr} + \gamma_8 a_{rkr} + \gamma_9 a_{rrk})\delta_{\ell m} +$$

$$+\gamma_{10} a_{k\ell m} + \gamma_{11} a_{km\ell} + \gamma_{12} a_{\ell km} + \gamma_{13} a_{\ell mk} + \gamma_{14} a_{mk\ell} + \gamma_{15} a_{m\ell k} .$$

A micro-fluid is a micropolar fluid if the gyration tensor $\underset{\sim}{\omega}$ is the angular velocity tensor for the particles, i.e. if $\omega_{ij} = -\omega_{ji}$, and if $\lambda^{ijk} = -\lambda^{ikj}$. In this case $a_{ijk} = -a_{jik}$.

The constitutive equations for micropolar fluids are much simpler than the equations for micro-fluids,

$$m_l^{\cdot k} = \alpha_v \omega^r_{,r}\delta^k{}_l + \beta_v \omega^k{}_{,l} + \gamma_v \omega_{l,m} g^{mk} ,$$

$$t^k_{\cdot l} = (-\pi + \lambda\omega^t_{,t})\delta^k{}_l + (2\mu_v + k_v) d^k_{\cdot l} + k_v(\Omega^k_{\cdot l} - \omega^k_{\cdot l}) \qquad (12.1.12)$$

where

$$\Omega_{ij} = v_{[i,j]} , \qquad \omega^t = \epsilon^{tij}\omega_{ij}$$

and the micro-stress average $\underset{\sim}{s}$ disappears from the equations. The spin becomes

$$\sigma_r = \epsilon_{rij}\sigma^{ij} = j\dot{\omega}_r . \qquad (12.1.13)$$

The equations of motion are

$$\rho(\dot{\underset{\sim}{v}} - \underset{\sim}{f}) = -\operatorname{grad}\pi + (\lambda_v + \mu_v)\operatorname{grad}\operatorname{div}\underset{\sim}{v} + (\mu_v + k_v)\Delta\underset{\sim}{v} + k_v(\underset{\sim}{\nabla}\times\underset{\sim}{\omega}), \qquad (12.1.14)$$

$$\rho(j\dot{\underset{\sim}{\omega}} - \underset{\sim}{l}) = (\alpha_v + \beta_v)\operatorname{grad}\operatorname{div}\underset{\sim}{\omega} + \gamma_v\Delta\underset{\sim}{\omega} + k_v(\underset{\sim}{\nabla}\times\underset{\sim}{v}) - 2k_v\underset{\sim}{\omega} . \qquad (12.1.15)$$

For $\underset{\sim}{\omega} = \underset{\sim}{l} = 0, k_v = \alpha_v = \beta_v = \gamma_v = 0$ these equations reduce to the Navier-Stokes equations. The theory of micropolar fluids includes four additional coefficients of viscosity, besides the two coefficients λ_v and μ_v which were known in the non-polar theory of viscous fluids.

The considered constitutive relations for micropolar fluids do not violate the Clausius-Duhem inequality, and the inequality only imposes certain restrictions on the coefficients of viscosity.

Aero, Bul'gin and Kuvshinskii [4] developed in 1964 independently a theory of fluids with the non-symmetric stress tensor, which is completely analogous to Eringen's theory of micropolar fluids, i.e. the directors represent rigid triads. Also in 1964 appeared a paper by Condif and Dahler [69] in which the fluid considered corresponds to the micropolar fluid, but their constitutive equations (linear) involve only five coefficients of viscosity. Allen, DeSilva and Kline [233] proposed a more general theory of fluids with deformable directors, but this theory is not completely developed. Recently appeared also a paper by Eringen [133] in which certain extensions of the theory of micro-fluids are studied in order to include deformable micro-elements. Quite recently also appeared a paper by Liu [272] in which some generalizations of the theory of micropolar fluids are suggested, in order to derive the equations for turbulent parallel flow from the general theory.

12.2 Dipolar Fluids and Fluids of Grade Two

Theory of dipolar fluids originates in the theory of multipolar continua proposed by Green and Rivlin [172, 173, 174] .

In the theory of dipolar fluids the constitutive equations are to be postulated (Bleustein and Green [38]), or derived (Plavsić [358-361]) for energy, entropy, heat flux, stress and dipolar stress, considering as constitutive variables the density of matter ϱ , gradients of the density $\varrho_{,i}$ and $\varrho_{,ij}$, temperature and temperature gradients θ , $\theta_{,i}$, $\theta_{,ij}$, and first and second gradients of the velocity, $v_{i,j}$, $v_{i,jk}$.

Assuming that the Helmholtz free energy function Ψ is a function of the form

$$\Psi = \Psi(\varrho, \varrho_{,i}, \varrho_{,ij}, d_{ij}, a_{ijk}, \theta, \theta_{,i}, \theta_{,ij}) \qquad (12.2.1)$$

where

$$d_{ij} = v_{(i,j)}, \quad a_{ijk} = v_{i,jk}, \qquad (12.2.2)$$

Bleustein and Green considered the Clausius-Duhem inequality in the form

$$-\varrho(\dot{\Psi} + \eta\dot{\theta}) - \frac{\theta_{,i} q^i}{\theta} + t^{ij} d_{ij} + \Sigma^{(ij)k} a_{kji} \geq 0 . \qquad (12.2.3)$$

Here $\Sigma^{(ij)k}$ are components of the "dipolar stress" which are symmetric in the first two indices. From an analysis it follows that Ψ cannot depend on other quantities, but on $\varrho, \varrho_{,i}$ and θ (cf. section 9.2 and equ. 9.2.11), and the inequality (12.2.3) reduces to ($\nu \equiv \delta^{mn}\varrho_{,m}\varrho_{,n}$):

$$[t_{ij} + \varrho^2 \frac{\partial\Psi}{\partial\varrho} g_{ij} + 2\varrho\frac{\partial\Psi}{\partial\nu}(\nu g_{ij} + \varrho_{,i}\varrho_{,j})]d^{ij} +$$

(12.2.4)

$$+ [\Sigma_{(ij)k} + \varrho^2\frac{\partial\Psi}{\partial\nu}(\varrho_{,i}g_{jk} + \varrho_{,j}g_{ik})]a^{kji} - \frac{\theta_{,i}q^i}{\theta} \geq 0 .$$

The constitutive equations are derived only for homogeneous incompressible fluids. For such a fluid we have $v^k_{,k} \equiv I_d = 0$, and Bleustein and Green obtained the following constitutive relations:

$$t_{ij} + \varphi g_{ij} = 2\mu d_{ij} + \beta\theta_{,ij} ,$$

$$\Sigma_{(ij)k} + \Psi_i\delta_{jk} + \Psi_j\delta_{ik} = h_1 g_{ij}a_{k\ell\ell} + h_2(a_{ijk} + a_{jik}) +$$

(12.2.5)

$$+ h_3 a_{kji} + \gamma g_{ij}\theta_{,k} ,$$

$$q_i = \alpha a_{ikk} + k\theta_{,i} .$$

Here φ , Ψ_i are some arbitrary functions to be determined in the course of solution of each particular problem.

Under certain, in the thermodynamical sense, more

restrictive conditions, Plavsić [361] derived the constitutive equations for dipolar fluids from Ziegler's principle of the least irreversible force (see section 8). He considered the dissipation function in the form

$$\varrho\Phi = \varrho\theta\dot{\eta} = t^{ij}d_{ij} + \Sigma^{(ij)k}a_{kij} . \qquad (12.2.6)$$

Since d_{ij} and a_{ijk} are objective tensors, the dissipation function may be regarded in the form

$$\Phi = \Phi(d_{ij}, a_{ijk}) , \qquad (12.2.7)$$

and from Ziegler's principle (8.41) follow the constitutive equations,

$$\begin{aligned} t^{(ij)} &= \varrho\left(\frac{\partial\Phi}{\partial d_{pq}}d_{pq} + \frac{\partial\Phi}{\partial a_{ijk}}a_{ijk}\right)^{-1}\Phi\frac{\partial\Phi}{\partial d_{ij}} , \\ \Sigma^{(ij)k} &= \varrho\left(\frac{\partial\Phi}{\partial d_{pq}}d_{pq} + \frac{\partial\Phi}{\partial a_{ijk}}a_{ijk}\right)^{-1}\Phi\frac{\partial\Phi}{\partial a_{kij}} . \end{aligned} \qquad (12.2.8)$$

These equations, when linearized, reduce to the equations (12.2.5).

In analogy to materials of grade two in the theory of elasticity, where the strain energy function is a function of the strain gradients, we may consider "fluids of grade two" where the dissipation function will depend on the second-order gradients of vorticity. This case was studied also by Plav sić [358, 360] ,

$$\Phi = \Phi(d_{ij}, w_{ijk}) . \tag{12.2.9}$$

Using again Ziegler's principle Plavsić obtained the constitutive equations for the symmetric part of the stress tensor and for the symmetric part $m^{i(jk)}$ of the couple-stress tensor, which is in complete analogy to the theory of elastic materials of grade two. When linearized, the constitutive equations read

$$\begin{aligned} t^{(ij)} &= -pg^{ij} + \eta_1 I_d g^{ij} + 2\eta_2 d^{ij} , \\ \mu_{ij} &= -(2\eta_3 w_{ij} + 2\eta_4 w_{ji}) , \end{aligned} \tag{12.2.10}$$

where μ_{ij} is the deviatoric part of the second-order couple--stress tensor. Here we have four coefficients of viscosity, but in the equations of motion besides the two coefficients which appear in the Navier-Stokes equations there will be present only one coefficient, the coefficient of "rotational viscosity". The equations of motion read

$$\varrho \dot{\underset{\sim}{v}} = -\operatorname{grad} p + \eta_2 \Delta \underset{\sim}{v} - \eta_3 \Delta\Delta \underset{\sim}{v} . \tag{12.2.11}$$

The essential difference between various approaches to polar continuum mechanics is in the assumed kinematics. For micropolar fluids there are two independent vectors which describe the configuration of a fluid, the velocity vector and the micro-rotation (or gyration) vector. In the theory of dipolar fluids and in the theory of fluids of grade two there is just one vector field, the velocity vector. However, since the second gradients of the velocity vector are objective quantities, their combinations contained in the vorticity gradients are also objective quantities and the dipolar fluids are a more general type of fluids than the fluids of grade two. Moreover, the theory of dipolar fluids by Bleustein and Green is based on a more general thermodynamical basis, valid also for heat-conducting fluids, and not on the restrictions as is in the case when we apply Ziegler's principle.

A very fine comparison of the theories of micropolar and dipolar fluids is made by Ariman [15] .

Independently of the difference, all existing theories of polar fluids predict certain effects which might be experimentally detected, and all theories are in agreement on the nature of these effects. In a number of papers the theories were applied to various flow problems, mostly to the study of channel and pipe flow and are obtained velocity profiles. Independently of the theory which was applied, the obtained velocities differ from the velocities obtained in the classical hydromechanics. Towards the middle of the channel or of the pipe the velocities are smaller in the case of polar fluids than in the case of a classical fluid. Plavsić [359] studied the viscometric flow of polar fluids and predicted theoretically certain measurable effects which might help in the determination of the coefficients of the rotational viscosity. It should be noted that already in 1962 S.C. Cowin [74] discovered that in oriented fluids such effects of rotational viscosity are to be expected.

The theory of Condif and Dahler [69] was inspired by the problem of fluids containing some rigid structures. The same problem was considered by Kirwan and Newman [233], who also considered fluids with deformable structures [234], basing their considerations on the theory of micropolar fluids. Afanas'ev and Nikolaevskii [7] considered the same problem as Kirwan and Newman in [233], but referring only to the work of Aero, Bul'gin and Kuvshinskii [4] and to Ericksen's papers on anisotropic fluids.

12.3 Liquid Crystals

In the section 7.2 we already derived the differential equations of motion of liquid crystals, according to Ericksen's theory. In addition to the contact and body forces which appear in (7.2.1), Leslie [268] introduced another force $\underset{\sim}{g}$ which is defined as an intrinsic director body force per unit volume. To avoid ambiguities in the notation, the director vector, which was previously denoted by $\underset{\sim}{d}$, we shall denote now by $\underset{\sim}{n}$. The equations of motion now read

$$\frac{d\varrho}{dt} + \varrho v^i_{,i} = 0 ,$$

$$\varrho \dot{v}^i = t^{ij}_{,j} + \varrho f^i , \qquad (12.3.1)$$

$$\varrho \ddot{n}^i = h^{ij}_{,j} + \varrho k^i + g^i .$$

With the aid of these equations the local energy balance law may be written in the form

$$\varrho \dot{\varepsilon} = \varrho h + q^k_{,k} + t^{ij} d_{ij} + h^{ij} N_{ji} - g^i N_i + \tilde{t}^{ij} w_{ij} , \qquad (12.3.2)$$

where

$$N_{ij} = \dot{n}_{i,j} + w_{ki} n^k_{,j} ; \quad N_i = \dot{n}_i + w_{ki} n^k , \qquad (12.3.3)$$

and

$$\tilde{t}^{ij} = t^{ij} - h^{jk} n^i_{,k} + g^j n^i . \qquad (12.3.4)$$

From the invariance of (12.3.2) under a superposed rigid rotation $\omega_{ij} = -\omega_{ji}$ it follows that $\tilde{t}^{ij}$ is symmetric.* This reduces (12.3.2) to

---

* This is not identically satisfied in (12.3.4) and has to be taken as a request into account when constitutive equations are formulated.

(12.3.5) $$\rho\dot{\varepsilon} = \rho h + q^{k}_{,k} + t^{ij}d_{ij} + h^{ij}N_{ji} - g^{i}N_{i} .$$

Leslie assumes the entropy inequality in the form

(12.3.6) $$\frac{d}{dt}\int_{v}\rho\eta\, dv - \int_{v}\frac{\rho h}{\theta}dv + \oint_{S}p^{i}ds_{i} \geq 0$$

where, according to some new concepts in thermodynamics, p_i is the entropy flux which is not necessarily equal to the heat flux per unit temperature. Writing $\varphi_i = q_i - \theta p_i$ and combining (12.3.5) and (12.3.6) we obtain

(12.3.7) $$t^{ij}d_{ij} + h^{ij}N_{ji} - g^{i}N_{i} - \theta_{,i}p^{i} - \rho(\dot{\Psi} + \eta\dot{\theta}) - \varphi^{i}_{,i} \geq 0 .$$

The quantities which have to be determined through the constitutive equations are

(12.3.8) $$\varepsilon, \eta, q^{i}, p^{i}, h^{ij}, t^{ij}, g^{i},$$

and the objective independent variables are

(12.3.9) $$\rho, \theta, n^{i}, n^{i}_{,j}, N_{i}, d_{ij}, \theta_{,i} .$$

From an analysis corresponding to that at the end of the section 9.2 we find that

(12.3.10) $$\Psi = \Psi(\rho, \theta, n^{i}, n^{i}_{,j}), \quad \eta = -\frac{\partial\Psi}{\partial\theta} .$$

For static isothermal deformations Leslie obtained

the following constitutive equations,

$$t^{ij} = -\varrho^2 \frac{\partial \Psi}{\partial \varrho} g^{ij} - \varrho g^{jl} \frac{\partial \Psi}{\partial n^k_{,i}} n^k_{,l} ,$$

$$h^i_{.j} = \varrho \frac{\partial \Psi}{\partial n^j_{,i}} + \alpha_0 D^i n_j , \qquad (12.3.11)$$

$$g_i = -\varrho \frac{\partial \Psi}{\partial n^i} - (\alpha_0 D^j n_i)_{,j} ,$$

where

$$D_i = n_i n^j_{,j} - n_j n^j_{,i} , \qquad (12.3.12)$$

and α_0 is a coefficient which is a scalar function of the temperature θ and of the magnitude of the director $\underset{\sim}{n}$. Ericksen [109] obtained the same constitutive equations, but without the terms involving the coefficient α_0.

Leslie's equations are applied to a number of special problems of interest to physicists working on liquid crystals. However, there is still a discrepancy between the theory and some observed phenomena. As is the case in the whole theory of polar media, the lack of estimates for constants which appear in the theory prevents a comparison of predicted results with the results of measurements.

13. Plasticity

The theory of plasticity represents even in the classical continuum mechanics a field in which certain fundamental problems are not solved. The existing engineering theories give for practical purposes sufficiently good results, but such theories represent only phenomenological descriptions which are more or less in good agreement with experiments, and the nature of the plastic flow from the physical standpoint, except in metals, is not well understood yet.

At a microscopic scale the mechanism of plastic flow in metals is explained as a consequence of the motion of dislocations, but there is still not existing a theory which is capable of connecting the phenomenological theories with dislocations. Polar media in the problem of plasticity play an interesting role, since one of the first applications of certain concepts in mechanics of Cosserat continua was just in the theory of dislocations (Günther [189]). However, the theories of plasticity in polar materials are still far from representing a missing link between the theory of dislocations and the problems of plastic flow.

In 1964 Komljenović [243] considered an elastic-plastic body with couple stresses. Assuming that the stress and couple-stress tensors may be separated into reversible (elastic) and irreversible (plastic)parts, he considered the energy balance equation,

$$\varrho\dot{\varepsilon} = \varrho\Theta\dot{\eta} + ({}_{E}t^{(ij)} + {}_{D}t^{(ij)})d_{ij} - ({}_{E}m^{ijk} + {}_{D}m^{ijk})w_{ij,k} \qquad (13.1)$$

and assumed that

$$\varepsilon = \varepsilon(\eta, x^{k}_{;K}, x^{k}_{;KL}) \qquad (13.2)$$

$$\Theta\dot{\eta} = \Phi(\dot{x}^{k}, \dot{x}^{k}_{;K}, \dot{x}^{k}_{;KL}) \qquad (13.3)$$

where Φ is the dissipation function.

For elastic parts of the stress and of the couple-stress tensor Komljenović obtained the well-known equations from the non-linear theory of materials of grade two. To obtain the constitutive equations for ${}_{E}\underset{\sim}{t}$ and ${}_{E}\underset{\sim}{m}$ he applied a method which corresponds to Ziegler's principle of least irreversible force. The yield condition is considered in the form

$$\Theta\dot{\eta} = \Phi \geqslant 0$$

i.e.

$$\Phi - k^{2} = 0 , \qquad (13.4)$$

where $k = \text{const}$. Since Φ is assumed in the form (13.3), only for linearized constitutive equations it was possible to substitute in Φ the rates $\dot{x}^{k}_{;K}$ and $\dot{x}^{k}_{;KL}$ by the stress and couple-stress tensors and to write the yield condition in the form

$$\Phi(\underset{\sim}{t}, \underset{\sim}{m}) - k^2 \geq 0 . \qquad (13.5)$$

For isotropic materials and in the absence of couple-stresses the dissipation function is an isotropic function and (13.5) reduces to the Henckey-Mises yield condition.

In 1967 Sawczuk [389] developed the theory of plastic flow in Cosserat continua with constrained rotations. The kinematical variables in Sawczuk's theory are

$$\begin{aligned} d_{ij} &= \dot{u}_{(i,j)} , \qquad x_{ij} = w_{i,j} \\ w^i &= \epsilon^{imn} w_{mn} = \epsilon^{imn} \dot{u}_{m,n} . \end{aligned} \qquad (13.6)$$

The dynamical variables are the symmetric stress tensor $\tau^{ij} = t^{(ij)}$ and the deviator of the couple-stress tensor $\mu_{ij} = m_{ij} - \frac{1}{3} m^k_{\cdot k} \delta_{ij}$. For the dissipation function it is assumed that it is of the form

$$\Phi = t^{ij} d_{ij} + \mu^{ij} x_{ij} \geq 0 . \qquad (13.7)$$

A further assumption is that the dynamical variables are homogeneous functions of degree zero in time and homogeneous of order zero in the kinematical variables,

$$\frac{\partial s_{ij}}{\partial d_{rs}} d_{rs} + \frac{\partial s_{ij}}{\partial x_{rs}} x_{rs} = 0 , \qquad (13.8a)$$

$$\frac{\partial \mu_{ij}}{\partial d_{rs}} d_{rs} + \frac{\partial \mu_{ij}}{\partial x_{rs}} x_{rs} = 0 \,, \tag{13.8b}$$

where s_{ij} is the deviatoric part of the stress tensor, $s_{ij} = \tau_{ij} - \frac{1}{3}\tau^{k}_{.k}\delta_{ij}$.

From (13.8) it follows that

$$s_{ij} = \alpha_A \overset{A}{T}_{ij} \,, \quad \mu_{ij} = \beta_B \overset{B}{T}_{ij} \,, \qquad A = 1,\dots,5 \,, \quad B = 1,\dots,8 \,, \tag{13.9}$$

where $\overset{A}{T}_{ij}$ and $\overset{B}{T}_{ij}$ are linearly independent tensorial functions of $\underset{\sim}{d}$ and $\underset{\sim}{x}$, and α_A and β_B are scalar functions of $\underset{\sim}{d}$ and $\underset{\sim}{x}$. In the tensorially linear form we have

$$s_{ij} = \alpha_1 d_{ij} + \alpha_2 x_{(ij)} \,, \quad \mu_{(ij)} = \beta_1 d_{ij} + \beta_2 x_{(ij)} \qquad \mu_{[ij]} = \gamma x_{[ij]} \,, \tag{13.10}$$

and α's, β's and γ are scalar functions of the second-order invariants of the kinematical variables.

Further analysis is based on the fact that in plasticity there does not exist a one-to-one correspondence between the invariants of the kinematical and of the dynamical variables. Since there is the same number of the variables $\underset{\sim}{s}$ and $\underset{\sim}{\mu}$ on one side, and $\underset{\sim}{d}$ and $\underset{\sim}{x}$ on the other side, from 13.9 it is possible to estab lish the relations between the two kinds of the invariants. The requirement that there are no 1:1 correspondences between the invariants yields the vanishing of the functional determinant. Denoting the invariants of the second order by

$$x = d_{ij} d_{ij} , \quad y = x_{(ij)} x_{(ij)} , \quad z = x_{[ij]} x_{[ij]} ,$$

$$\xi = s_{ij} s_{ij} , \quad \eta = \mu_{(ij)} \mu_{(ij)} , \quad \zeta = \mu_{[ij]} \mu_{[ij]}$$

etc., we have

$$\text{(13.11)} \qquad \left| \frac{\partial(x, y, z)}{\partial(\xi, \eta, \zeta)} \right| = 0 .$$

The tensorially linear flow law may be considered in the form

$$\text{(13.12)} \quad d_{ij} = \frac{\xi}{A} s_{ij} , \quad x_{(ij)} = \frac{\xi}{A\lambda_1^2} \mu_{(ij)} , \quad x_{[ij]} = \frac{\xi}{A\lambda_2^2} \mu_{[ij]} ,$$

where

$$\text{(13.13)} \qquad \frac{y}{x} = \frac{B}{A} \frac{\eta}{\xi} = \lambda_1^2 \frac{\eta}{\xi} , \quad \frac{z}{x} = \lambda_2^2 \frac{\zeta}{\xi} ,$$

and λ_1 and λ_2 have the dimension of length. Elimination of S from the relations between the invariants x, y, z leads to the establishment of the yield condition. (13.12) is in general not compatible with any potential rule for plastic flow.

Lippamn [270] considered a Cosserat continuum with directors which represent rigid triads. The kinematical variables of Lippman's theory are

$$d_{ij} = \dot{e}_{ij} = \partial_{(i}v_{j)}\,, \quad w_{ij} = \partial_{[i}v_{j]}\,, \quad x_{ij} = \partial_i \omega_j\,, \qquad (13.14)$$

and the dynamic variables are

$$t^{ij} \neq t^{ji}, \quad m^{ij} \neq m^{ji}. \qquad (13.15)$$

The dynamic variables represent a system of 18 components of a generalized force $\underset{\sim}{Q} = \{Q_1,\ldots,Q_{18}\}$. 18 components of d_{ij}, x_{ij} and of $\underset{\sim}{\Omega} = \underset{\sim}{\omega} - \underset{\sim}{x}$ are considered as 18 components of a generalized velocity $\underset{\sim}{q} = \{q_1,\cdots,q_{18}\}$.

The basic assumption of the theory is the extremum principle of Sadowski, Philips and Hill: for arbitrary velocities $\underset{\sim}{q}$ the forces $\underset{\sim}{Q}$ have such values that the shape-deformation action

$$\Lambda = \underset{\sim}{Q}\cdot\underset{\sim}{q} \qquad (13.16)$$

is maximal, $\delta\Lambda = 0$.

The most interesting assumption of Lippmann is that there are at least 2 and at most 18 yield-conditions,

$$fp(\underset{\sim}{Q}) = 0\,, \quad (2 \leqslant p \leqslant 18)\,. \qquad (13.17)$$

Then we have simultaneously

$$(13.18) \qquad q_k \delta Q_k = 0 , \quad \frac{\partial f_p}{\partial Q_k} \delta Q_k = 0 ,$$

and consequently

$$(13.19) \qquad q_k = \lambda_p \frac{\partial f_p}{\partial Q_k} ,$$

where λ_p are proportionality factors.

For various specific conditions Lippmann derived various yield conditions of the classical theory of plasticity as special cases of his theory. He also applied the theory to a number of problems which are of technical and practical importance.

At the end we shall mention here also that some attempts were recently made for the formulation of various theories of other anelastic phenomena. There are papers on visco-elasto-plasticity (Misicu [294]), and on visco-plasticity (Radenković and Plavsić [362]), as well as on viscoelasticity (e.g. Eringen [129] , Askar, Cakmak and Ariman [20] , DeSilva and Kline [83] , McCarthy and Eringen [278], etc.). All these theories represent very important contributions which we unfortunately have no time to analyze in detail here, but as a general conclusion we might say that even in the polar theories of elasticity, and elasticity is physically the simplest situation, we have not succeeded yet in establishing a general theory and that in the theories involving irreversible phenomena a great deal of work remains to be done.

Appendix

For theoretical considerations it seems to me that the most suitable in the nonlinearized expositions is the notation of the double tensor field theory (cf. Ericksen "Tensor Fields" [100]). Assuming that the readers are familiar with the tensor analysis, the aim of this Appendix is to present only a survey of notation and some basic properties of ordinary and double tensor fields which are used in the lectures.

A1. Coordinates. Tensors.

An ordered set of numbers $\underset{\sim}{x}=\{x^1,x^2,\dots,x^n\}$ (we consider only real numbers) represents an arithmetic point. The numbers x^k are coordinates of the point $\underset{\sim}{x}$. The set of all possible arithmetic points, obtained when the coordinates take all possible values, represents an n-dimensional arithmetic space A_n.

If M is a set of objects m , such that there is a 1:1 correspondence between the objects of the set M and the points $\underset{\sim}{x}$ of a region A of A_n we may say that the numbers x^i are coordinates of the objects m , and that the objects m are pictures of the arithmetic points $\underset{\sim}{x}$.

If there is a 1:1 mapping of points $\underset{\sim}{x}$ of a region A in A_n upon points $\underset{\sim}{\bar{x}}$ of a region $\bar{A}$ in the same A_n ,

$$
\text{(A1.1)} \qquad \begin{aligned} x^k &= x^k(\bar{x}^1, \bar{x}^2, \ldots, \bar{x}^n)\,, \\ \bar{x}^k &= \bar{x}^k(x^1, x^2, \ldots, x^n)\,, \end{aligned}
$$

we say that the $\bar{x}^k$ represent another coordinate system with respect to which the objects m are determined. The set M of objects m, together with the coordinate system x^k, and a group of transformations (A1.1) which introduces all admissible systems, represents an n-dimensional geometric space X_n. The objects m are now points of the space X_n.

The coordinate transformations are transformations of numbers characterizing the same point m.

If R is a region in X_n with points $A \in R$ referred to a coordinate system x^k, and if $\bar{R}$ is another region in X_n with points B referred to a system of coordinates X^K, the 1:1 mappings of the points of R upon the points of $\bar{R}$,

$$
\text{(A1.2)} \qquad \begin{aligned} x^k_A &= x^k(X^1_B, \ldots, X^n_B)\,, \\ X^K_B &= X^K(x^1_A, \ldots, x^n_A)\,, \end{aligned}
$$

represent a point transformation.

In the following, if x^k are coordinates of a point in X_n, we say it is the point $\underset{\sim}{x}$.

A geometric quantity in X_n at a point $\underset{\sim}{x}$ is defined by a set of numbers, say N, and by a transformation law which enables us to determine these numbers when a coordinate transformation is performed. If x_P^k and $\bar{x}_P^k$ are coordinates of a point P in X_n given with respect to two coordinate systems, and F_Ω, $\Omega=1,2,\dots,N$ are the components of a geometric object $\underset{\sim}{F}$, the general transformation law has the form

$$F_\Omega\{\underset{\sim}{\bar{x}}_P\} = \Phi_\Omega\left(F_1\{\underset{\sim}{x}_P\},\dots,F_N\{\underset{\sim}{x}_P\},\underset{\sim}{x}_P,\underset{\sim}{\bar{x}}_P,\frac{\partial\bar{x}^k}{\partial x^m},\dots,\frac{\partial^q\bar{x}^k}{\partial x^{m_1}\dots\partial x^{m_q}},\dots\right).$$

If the transformation law does not depend explicitly on the coordinates of the point P, and on the partial derivatives of higher order than the first, the geometric object is a geometric quantity.

A scalar is a geometric quantity with one component and with the transformation law

$$\varphi(x^1,\dots,x^n) = \varphi(\bar{x}^1,\dots,\bar{x}^n). \tag{A1.3}$$

Covariant vectors are quantities with the number of components equal to the number of the dimensions of the space, $n=N$. If v_k and $\bar{v}_l$ are components of a covariant vector $\underset{\sim}{v}$ at a point $\underset{\sim}{x}$, the transformation law for covariant vectors reads

$$\bar{v}_l = v_k\frac{\partial x^k}{\partial\bar{x}^l}, \tag{A1.4a}$$

(A1.4b) $$v_k = \bar{v}_l \frac{\partial \bar{x}^l}{\partial x^k} \qquad (k,l = 1,2,\ldots,n) .$$

Here and in the following we apply the usual summation convention for repeated indices.

For a <u>contravariant vector</u> $\underset{\sim}{w}$ with components w^k and $\bar{w}^l$ the transformation law reads

(A1.5) $$\bar{w}^l = w^k \frac{\partial \bar{x}^l}{\partial x^k} , \qquad w^k = \bar{w}^l \frac{\partial x^k}{\partial \bar{x}^l} .$$

A tensor $\underset{\sim}{T}$ of covariant order p and contravariant order q is a quantity with n^{p+q} components $T_{i_1 \ldots i_p}^{\cdots \; j_1 \ldots j_q}$ and with the transformation law

(A1.6) $$\bar{T}_{i_1 \ldots i_p}^{\cdots\cdots j_1 \ldots j_q} = T_{k_1 \ldots k_p}^{\cdots\cdots l_1 \ldots l_q} \frac{\partial x^{k_1}}{\partial \bar{x}^{i_1}} \cdots \frac{\partial x^{k_p}}{\partial \bar{x}^{i_p}} \frac{\partial \bar{x}^{j_1}}{\partial x^{l_1}} \cdots \frac{\partial \bar{x}^{j_q}}{\partial x^{l_q}} .$$

The order of this tensor is $p + q$.

A tensor all of whose indices are superscripts (subscripts) is said to be a contravariant (covariant) tensor.

If the components of a tensor remain unchanged when two of its co- or contravariant indices interchange their places, we say that the tensor is symmetric with respect to these indices, e.g.

$$T_{ijkl} = T_{ikjl}\,, \qquad T^{pqr} = T^{qpr}\,.$$

If components of a tensor change sign when two of its co- or contravariant indices interchange their positions, the tensor is antisymmetric, e.g.

$$T_{ijkl} = -T_{ikjl}\,, \qquad T^{pqr} = -T^{qpr}\,.$$

A second-order tensor may always be decomposed into its symmetric part,

$$T_{(ij)} = \frac{1}{2}(T_{ij} + T_{ji})\,, \qquad T^{(ij)} = \frac{1}{2}(T^{ij} + T^{ji})\,, \tag{A1.7}$$

and into its antisymmetric part,

$$T^{[ij]} = \frac{1}{2}(T^{ij} - T^{ji})\,, \qquad T_{[ij]} = \frac{1}{2}(T_{ij} - T_{ji})\,, \tag{A1.8}$$

such that

$$T^{ij} = T^{(ij)} + T^{[ij]}\,, \qquad T_{ij} = T_{(ij)} + T_{[ij]}\,. \tag{A1.9}$$

There are tensors defined simultaneously with respect to two points of the space, and these two points are, in general, referred to two different coordinate systems, say x^k and X^K. Such tensors represent the <u>double tensor fields</u>. Let $t^k_{.K}(\underset{\sim}{x},\underset{\sim}{X})$ be such a tensor. With respect to coordinate transformations at $\underset{\sim}{x}$ it transforms like a contravariant vector, and with respect to coordinate transformations at $\underset{\sim}{X}$ it transforms like a covariant vector,

$$\bar{t}^{\ell}_{.L} = t^k_{.K} \frac{\partial \bar{x}^{\ell}}{\partial x^k} \frac{\partial X^K}{\partial \bar{X}^L}. \tag{A1.10}$$

Further examples of the double tensor fields are partial derivatives of the point transformations (A1.2)

$$F^k_{.K} \equiv \frac{\partial x^k}{\partial X^K} \equiv x^k_{;K}\,; \qquad F^{.K}_{k} \equiv \frac{\partial X^K}{\partial x^k} \equiv X^K_{;k}. \tag{A1.11}$$

In Euclidean spaces there exist rectilinear orthogonal (Cartesian) systems z^α, $\alpha = 1,2,\ldots,n$, and if such a coordinate system is admissible in an X_n, besides some other properties which will be mentioned later, we say that it is Euclidean space. The unit vectors in the directions of the coordinate lines z^α we shall denote by $\underset{\sim}{e}^\alpha = \underset{\sim}{e}_\alpha$. The position of a point $\underset{\sim}{z}$ in E_n is determined by the position vector $\underset{\sim}{r}$,

$$\underset{\sim}{r} = z^\alpha \underset{\sim}{e}_\alpha, \tag{A1.12}$$

where $r^\alpha = z^\alpha$ are the components of $\underset{\sim}{r}$. If x^i is an admissible coordinate system in Euclidean space, i.e. if there exist the coor-

the coordinate transformations

$$x^i = x^i(z^1,\ldots,z^n)$$
$$z^\alpha = z^\alpha(x^1,\ldots,x^n) \qquad \text{(A1.13)}$$

which are analytic functions in the neighbourhood of the point $\underset{\sim}{z}$, the components of the position vector $\underset{\sim}{r}$ with respect to the system x^i are given by

$$r^i = z^\alpha \frac{\partial x^i}{\partial z^\alpha} ,$$
$$z^\alpha = r^i \frac{\partial z^\alpha}{\partial x^i} . \qquad \text{(A1.14)}$$

Denoting by $\underset{\sim}{g}_i$ the base vectors of the coordinate system x^i, the position vector $\underset{\sim}{r}$ may be expressed now in the form

$$\underset{\sim}{r} = r^i \underset{\sim}{g}_i , \qquad \text{(A1.15)}$$

where

$$\underset{\sim}{g}_i = \underset{\sim}{e}_\alpha \frac{\partial z^\alpha}{\partial x_i} = \frac{\partial \underset{\sim}{r}}{\partial x^i} . \qquad \text{(A1.16)}$$

The reciprocal base vectors $\underset{\sim}{g}^i$, defined by the relations

$$\underset{\sim}{g}^i = \underset{\sim}{e}^\alpha \frac{\partial x^i}{\partial z^\alpha} \qquad \text{(A1.17)}$$

represent the reciprocal vectorial base. For Cartesian coordinates, the scalar products of the base vectors are

(A1.18) $$\underset{\sim}{e}^{\alpha}\underset{\sim}{e}_{\beta} = \delta^{\alpha}_{\beta}, \quad \underset{\sim}{e}_{\alpha}\underset{\sim}{e}_{\beta} = \delta_{\alpha\beta}, \quad \underset{\sim}{e}^{\alpha}\underset{\sim}{e}^{\beta} = \delta^{\alpha\beta},$$

where $\delta^{\alpha}_{\beta} = \delta_{\alpha\beta} = \delta^{\alpha\beta} = \begin{cases} 1, \alpha = \beta \\ 0, \alpha \neq \beta \end{cases}$ are the Kronecker symbols. Hence

$$\underset{\sim}{g}^{i}\underset{\sim}{g}_{j} = \underset{\sim}{e}^{\alpha}\underset{\sim}{e}_{\beta}\frac{\partial x^{i}}{\partial z^{\alpha}}\frac{\partial z^{\beta}}{\partial x^{j}} = \delta^{\alpha}_{\beta}\frac{\partial x^{i}}{\partial z^{\alpha}}\frac{\partial z^{\beta}}{\partial x^{j}} = \delta^{i}_{j}.$$

We shall use the symbol $\underset{\sim}{1}$ for the matrix $\{\delta^{\alpha}_{\beta}\}$.

The scalar products of the base vectors $\underset{\sim}{g}_{i}$ and $\underset{\sim}{g}^{i}$ give the components of the fundamental tensor (g_{ij} and g^{ij}) for the systems of coordinates x^{i}, which is a symmetric tensor,

(A1.19) $$g_{ij} = \underset{\sim}{g}_{i}\underset{\sim}{g}_{j} = g_{ji} = \delta_{\alpha\beta}\frac{\partial z^{\alpha}}{\partial x^{i}}\frac{\partial z^{\beta}}{\partial x^{j}},$$

and also

(A1.20) $$g^{ij} = \underset{\sim}{g}^{i}\underset{\sim}{g}^{j} = g^{ji} = \delta^{\alpha\beta}\frac{\partial x^{i}}{\partial z^{\alpha}}\frac{\partial x^{j}}{\partial z^{\beta}}.$$

Transvection of co- and contravariant components of the fundamental tensor gives the components of the unit tensor,

(A1.21) $$g^{ij}g_{jk} = \delta^{i}_{k}.$$

Denoting by G^{ij} the cofactor in the determinant $g = \det g_{ij}$, corresponding to the element g_{ji} such that

(A1.22) $$g\delta^{j}_{k} = G^{ji}g_{ik},$$

from (A1.22) we have

$$g^{ji} = \frac{G^{ji}}{g}, \quad g_{ji} = g\,G_{ji}, \tag{A1.23}$$

where G_{ji} is the cofactor in $\det g^{ij}$ corresponding to the element g^{ij}, and

$$\det g^{ij} = (\det g_{ij})^{-1}. \tag{A1.24}$$

In 3-dimensional Euclidean spaces the vectorial product of two base vectors $\underset{\sim}{e}_\alpha$ and $\underset{\sim}{e}_\beta$, $\alpha \neq \beta$ is the vector $\pm\underset{\sim}{e}_\gamma$, α, β, γ all different. If $\alpha\,\beta\,\gamma$ is an even permutation of the numbers 123, we have

$$\underset{\sim}{e}_\alpha \times \underset{\sim}{e}_\beta = \underset{\sim}{e}_\gamma \qquad (\alpha, \beta, \gamma \neq) \tag{A1.25}$$

and if it is an odd permutation,

$$\underset{\sim}{e}_\alpha \times \underset{\sim}{e}_\beta = -\underset{\sim}{e}_\gamma \qquad (\alpha, \beta, \gamma \neq). \tag{A1.26}$$

Hence we may define completely antisymmetric unit tensors $e_{\alpha\beta\gamma}$ and $e^{\alpha\beta\gamma}$ by the scalar products

$$\begin{aligned} (\underset{\sim}{e}_\alpha \times \underset{\sim}{e}_\beta)\,\underset{\sim}{e}_\gamma &= e_{\alpha\beta\gamma} \\ (\underset{\sim}{e}^\alpha \times \underset{\sim}{e}^\beta)\,\underset{\sim}{e}^\gamma &= e^{\alpha\beta\gamma}. \end{aligned} \tag{A1.27}$$

Under arbitrary coordinate transformations the unit tensors $\underset{\sim}{e}$ do not behave as tensors. The transformation law involves the Jacobian of the coordinate transformation and such tensors are named <u>relative tensors.</u> However, if we make the scal ar products analogous to (A1.27), we obtain using the relations (A1.16, 17)

$$(A1.28)\qquad (g_i \times g_j)\underset{\sim}{g}_k = e_{\alpha\beta\gamma}\frac{\partial z^\alpha}{\partial x^i}\frac{\partial z^\beta}{\partial x^j}\frac{\partial z^\gamma}{\partial x^k} = \left(\det\frac{\partial z^\lambda}{\partial x^l}\right)e_{ijk} ,$$

where e_{ijk} are now numerical symbols with the same meaning the unit tensors for Cartesian coordinates have. From (A1.19) we have now

$$g = (\det\delta_{\alpha\beta})\left(\det\frac{\partial z^\alpha}{\partial x^i}\right)^2 = \left(\det\frac{\partial z^\alpha}{\partial x^i}\right)^2 ,$$

and therefore for (A1.28) we may write

$$(A1.29)\qquad \varepsilon_{ijk} \equiv (\underset{\sim}{g}_i \times \underset{\sim}{g}_j)\underset{\sim}{g}_k .$$

Similarly

$$(A1.30)\qquad \varepsilon^{ijk} \equiv (g^i \times g^j)g^k = \frac{1}{\sqrt{g}}e^{ijk} .$$

The quantities ε_{ijk} and ε^{ijk} are true tensors under arbitrary coordinate transformations and often they are referred to as the Ricci tensors.

Using Ricci tensors an antisymmetric tensor may be represented by a vector. For instance, if $M^{ij} = -M^{ij}$, the tensor $\underset{\sim}{M}$ has three independent nonvanishing components in E_3 and we may represent it by a covariant vector

$$M_i = \frac{1}{2}\varepsilon_{ijk}M^{jk} , \qquad (A1.31)$$

$$(M^{jk} = \varepsilon^{ijk}M_i) .$$

Analogously, if $m^{ijk} = -m^{jik}$ is an antisymmetric third -order tensor, we may represent it as a second- order mixed tensor,

$$m_l^{\ k} = \frac{1}{2}\varepsilon_{lij}\, m^{ijk} , \qquad (A1.32)$$

$$(m^{ijk} = \varepsilon^{lij} m_l^{\ k}) .$$

Using the components of the fundamental tensor the operation of raising and lowering of indices may be defined, such that

$$g_{ij}\, T^{\cdots i \cdots}_{\cdots\cdots\cdots} = T^{\cdots\cdots\cdots}_{\cdots j \cdots} , \qquad (A1.33)$$

and

$$g^{ij}\, T^{\cdots\cdots\cdots}_{\cdots j \cdots} = T^{\cdots i \cdots}_{\cdots\cdots\cdots} . \qquad (A1.34)$$

Thus

$$t^{ij} = g^{jk} t^{i}_{.k} = g^{jk} g^{il} t_{lk} = g^{il} t_l^{\ .j} ,$$

and for the scalar product of two vectors, say $\underset{\sim}{u}$ and $\underset{\sim}{v}$, we may write

$$\text{(A1.35)} \qquad \underset{\sim}{u}\cdot\underset{\sim}{v} = u^i v_i = g_{ij}u^i v^j = g^{ij}u_i v_j = u_i v^i .$$

The vectorial product of two vectors, say $\underset{\sim}{a}$ and $\underset{\sim}{b}$, is a second-order antisymmetric tensor,

$$\text{(A1.36)} \qquad \begin{aligned} \underset{\sim}{a}\times\underset{\sim}{b} &= \{a^i b^j - a^j b^i\} = \{c^{ij}\} , \\ c^{ij} &= -c^{ji} , \end{aligned}$$

and using the Ricci tensor we may represent it as a vector $\underset{\sim}{c}$,

$$\text{(A1.37)} \qquad c_k = \frac{1}{2}\varepsilon_{ijk}c^{ij} = \varepsilon_{ijk}a^i b^j .$$

Tensors, as geometrical quantities, are defined at points of the space, and the operations of addition may be performed only if the tensors considered are brought to the same point of the space. If we have to add two tensors, or to compare them, and they are not defined at the same point, one of the tensors must be shifted parallely to the point in which the other tensor is defined. In Cartesian coordinates the components of a vector which represents a field of parallel vectors at all points of the space are equal, but with respect to curvilinear coordinates this is not true and we have to define the operation of parallel shifting which will enable us to compare components of tensors which are not given at the same point.

Let $\underset{\sim}{v}$ be a field of parallel vectors in E_3 and let

v^k be its components at a point $\underset{\sim}{x}$, and V^K its components at a point $\underset{\sim}{X}$. The two points may, in general, be determined with respect to two different coordinate systems, x^k and X^K. Let $\underset{\sim}{z}$ and $\underset{\sim}{Z}$ be the coordinates of the two points considered with respect to an absolute Cartesian system of reference and v^λ and V^Λ the components of the vector field $\underset{\sim}{v}$ with respect to this Cartesian system. Since by assumption $\underset{\sim}{v}$ is a field of parallel vectors, we have

$$v^\lambda = \delta^\lambda_\Lambda V^\Lambda , \quad \text{or} \quad V^\Lambda = \delta^\Lambda_\lambda v^\lambda . \tag{A1.38}$$

According to the transformation law for vectors we have

$$v^\lambda = v^k \frac{\partial z^k}{\partial x^\lambda} , \qquad V^\Lambda = V^K \frac{\partial Z^\Lambda}{\partial X^k} , \tag{A1.39}$$

and the relations (A1.38) may be written in the form

$$v^k = \delta^\lambda_\Lambda \frac{\partial x^k}{\partial z^\lambda} \frac{\partial Z^\Lambda}{\partial X^K} V^K , \qquad V^K = \delta^\Lambda_\lambda \frac{\partial X^K}{\partial Z^\Lambda} \frac{\partial z^\lambda}{\partial x^k} v^k . \tag{A1.40}$$

The quantities

$$g^k_{.K} = \delta^\lambda_\Lambda \frac{\partial x^k}{\partial z^\lambda} \frac{\partial Z^\Lambda}{\partial X^K} , \qquad g_k^{.K} = \delta^\Lambda_\lambda \frac{\partial X^K}{\partial Z^\Lambda} \frac{\partial z^\lambda}{\partial x^k} ,$$
$$(\text{with } g^k_{.K} g_l^{.K} = \delta^k_l , \quad g^k_{.K} g_k^{.L} = \delta^L_K) , \tag{A1.41}$$

are the <u>Euclidean shifters</u> (Doyle and Ericksen [92] , Toupin [460]). Using the shifters we may perform the shifting of an arbitrary tensor from one point of the space to another.

As an example let us consider a vector field $\underset{\sim}{v}$ at a point (R,Φ) given with respect to a system of polar coordinates in the Euclidean plane, and let us shift it to a point (r,φ) given with respect to the same system of coordinates. Since $Z^1=X$, $Z^2=Y$; $z^1=x$, $z^2=y$; $X^1=R$, $X^2=\Phi$; $x^1=r$, $x^2=\varphi$ and since the coordinate transformations at the two considered points are

$$X = R\cos\Phi\,, \qquad Y = R\sin\Phi$$

$$x = r\cos\varphi\,, \qquad y = r\sin\varphi$$

from (A1.41) we obtain the following expressions for the components of the shifter:

$$g^1_{.1} = \cos(\varphi-\Phi)\,, \qquad g^1_{.2} = R\sin(\varphi-\Phi)$$

$$g^2_{.1} = \frac{1}{r}\sin(\Phi-\varphi)\,, \qquad g^2_{.2} = \frac{R}{r}\cos(\Phi-\varphi)\,.$$

Using now $(A1.40)_1$ we easily obtain the components v^k of the vector $\underset{\sim}{v}$ when shifted from the point (R,Φ) to the point (r,φ):

$$v^1 = V^1\cos(\varphi-\Phi) + RV^2\sin(\varphi-\Phi)\,,$$

$$v^2 = \frac{1}{r}V^1\sin(\Phi-\varphi) + \frac{R}{r}V^2\cos(\Phi-\varphi)\,.$$

The shifters $g^{k}_{.K}$ represent another example of double tensors, and applying them to an arbitrary tensor by parallel shifting we perform the <u>conversion of indices</u>, e.g.

$$g^{\ell}_{.K} T^{K}_{.PQ} = T^{\ell}_{.PQ} .$$

If g_{mn} and G_{MN} are components of the fundamental tensors corresponding to the coordinate systems x^k and X^K at the points $\underset{\sim}{x}$ and $\underset{\sim}{X}$ of the space, from (A1.19) and (A1.41) we obtain

$$g^{k}_{.K} g^{\ell}_{.L} g_{k\ell} = \delta_{AB} \frac{\partial Z^A}{\partial X^K} \frac{\partial Z^B}{\partial X^L} = G_{KL} .$$

Let $\underset{\sim}{g}_k$, $\underset{\sim}{g}^k$, $\underset{\sim}{G}_K$ and $\underset{\sim}{G}^K$ be base vectors for curvilinear coordinate systems x^k and X^K respectively. According to (A1.16, 17) we have

$$\underset{\sim}{g}_k = \frac{\partial \underset{\sim}{r}}{\partial x^k} = \frac{\partial z^\alpha}{\partial x^k} \underset{\sim}{e}_\alpha , \quad \underset{\sim}{g}^k = \frac{\partial x^k}{\partial z^\alpha} \underset{\sim}{e}^\alpha ,$$

$$\underset{\sim}{G}_K = \frac{\partial \underset{\sim}{R}}{\partial X^K} = \frac{\partial Z^\alpha}{\partial X^K} \underset{\sim}{e}_\alpha , \quad \underset{\sim}{G}^K = \frac{\partial X^K}{\partial Z^\alpha} \underset{\sim}{e}^\alpha .$$

The Euclidean shifters may be defined as scalar products of the base vectors considered at two different points of the space,

$$\underset{\sim}{g}_k \underset{\sim}{G}^K = g_k^{\cdot K} , \quad \underset{\sim}{g}^k \underset{\sim}{G}_K = g^{k}_{.K} , \tag{A1.42}$$

and we may write the following formulae:

$$\text{(A1.43)} \qquad G_{KL}\, g_k^{.K} = g_{kL} = \underset{\sim}{g}_k \underset{\sim}{G}_L\,, \qquad g_{K\ell}\, g_{.K}^{\ell} = g_{kK} = \underset{\sim}{g}_k \underset{\sim}{G}_K\,.$$

The infinitesimal displacements $d\underset{\sim}{r}$ at a point $\underset{\sim}{x}$ are vectors of the form

$$\text{(A1.44)} \qquad d\underset{\sim}{r} = dx^i \underset{\sim}{g}_i = \frac{\partial z^\alpha}{\partial x^i} dx^i \underset{\sim}{e}_\alpha\,,$$

and the square of the displacement $d\underset{\sim}{r}$ represents the fundamental (metric) form for the space and for the considered system of coordinates,

$$\text{(A1.45)} \qquad ds^2 = d\underset{\sim}{r}\, d\underset{\sim}{r} = \underset{\sim}{g}_i \underset{\sim}{g}_j\, dx^i dx^j = g_{ij}\, dx^i dx^j\,.$$

Hence, the fundamental tensor in the Euclidean space is the metric tensor.

Physical components of vectors and tensors are defined only for orthogonal systems of coordinates ($g_{ij} = 0$ for $i \neq j$) . If we write for the base vectors $\underset{\sim}{g}_i = h_i \underset{\sim}{g}_{0i}$, with $h_i = |\underset{\sim}{g}_i|$, where g_{0i} are unit vectors colinear with the base vectors, evidently we have

$$\text{(A1.46)} \qquad h_i = \sqrt{g_{ii}}$$

and

$$\underset{\sim}{g}_{0i} = \frac{\underset{\sim}{g}_i}{\sqrt{g_{ii}}} \quad \text{(not summed)}. \tag{A1.47}$$

We may also write $\underset{\sim}{g}^i = h^i \underset{\sim}{g}_0^i$ with

$$h^i = \sqrt{g^{ii}}, \quad \underset{\sim}{g}_0^i = \frac{1}{\sqrt{g^{ii}}}\underset{\sim}{g}^i \quad \text{(not summed)}. \tag{A1.48}$$

and from (A1.23) we see that for orthogonal coordinate systems

$$g^{ii} = \frac{1}{g^{ii}} . \tag{A1.49}$$

The physical components of a vector are scalar products of the vector and of unit vectors colinear with the base vectors. Thus, for the physical components of a vector $\underset{\sim}{V}$ which will be denoted by $V(i)$ since the indices are neither co-, nor contravariant we have

$$\begin{aligned} V(i) &= \underset{\sim}{V}\underset{\sim}{g}_{0i} = \frac{1}{\sqrt{g_{ii}}} V^k \underset{\sim}{g}_k \underset{\sim}{g}_i = V_i/\sqrt{g_{ii}} = \\ &= \underset{\sim}{V}\underset{\sim}{g}_0^i = V^i/\sqrt{g^{ii}} . \end{aligned} \tag{A1.50}$$

Physical components of tensors are defined in analogy to the definition just introduced for vectors, e.g. for a second-order tensor we have

$$t(ij) = \frac{t^{ij}}{\sqrt{g^{ii}g^{jj}}} = \frac{t_{ij}}{\sqrt{g_{ii}g_{jj}}} = \frac{t^i_{\cdot j}}{\sqrt{g^{ii}g_{jj}}} . \tag{A1.51}$$

Besides the decomposition of a second-order tensor into its symmetric and antisymmetric parts, for mixed tensors also may be introduced a decomposition into its deviatoric and spherical parts. The <u>deviator</u> of a tensor $\underset{\sim}{T}$ is defined by the expression

(A1.52) $$ {}^{D}T^{i}_{\cdot j} = T^{i}_{\cdot j} - \frac{1}{3}T^{k}_{\cdot k}\delta^{i}_{j} , $$

and its <u>spherical tensor</u> will be

$$ {}^{S}T^{i}_{\cdot j} = \frac{1}{3}T^{k}_{\cdot k}\delta^{i}_{j} , $$

such that for the considered tensor we have

$$ T^{i}_{\cdot j} = {}^{D}T^{i}_{\cdot j} + {}^{S}T^{i}_{\cdot j} . $$

A2. Invariants

Let $\underset{\sim}{T}_{(1)}, \ldots, \underset{\sim}{T}_{(K)}$ be tensor variables. Any scalar function of these variables,

(A2.1) $$ f(\underset{\sim}{T}_{(1)}, \ldots, \underset{\sim}{T}_{(k)}) , $$

which remains invariant with respect to arbitrary coordinate transformations is an absolute <u>invariant</u> of the tensor $\underset{\sim}{T}_{(1)},\ldots,\underset{\sim}{T}_{(K)}$. However, there are invariants only with respect to some particular groups of transformations. We are mostly interested in ortho gonal transformations.

For a linear transformation of Cartesian coordinates

$$\bar{z}^{\lambda} = Q^{\lambda}_{\cdot\mu} z^{\mu} + a^{\lambda}, \qquad z^{\lambda} = Q_{\mu}^{\cdot\lambda} \bar{z}^{\mu} + b^{\lambda}, \tag{A2.2}$$

we say that it is orthogonal if

$$Q \equiv \det Q^{\lambda}_{\cdot\mu} = \pm 1, \tag{A2.3}$$

and the matrix of the coefficients of this transformation has the properties $\underset{\sim}{Q}^{T} = \underset{\sim}{Q}^{-1}$, where T denotes the transposition of a matrix. If $Q = \pm 1$, the transformation (A2.2) belongs to the group of full orthogonal transformations, and if $Q = +1$, we have the group of proper transformations.

Functions (A2.1) invariant with respect to the full orthogonal group are called isotropic invariants, and if they are invariant only with respect to a subgroup of the full orthogonal group, then it is said that they are relative invariants with respect to that subgroup. If a function is invariant only under the transformations of the group of proper orthogonal transformations, such invariants are called hemitropic invariants.

If $\underset{\sim}{T}$ is a symmetric tensor of the second order, the principal invariants of $\underset{\sim}{T}$ are:

$$I_T = \frac{1}{1!} \delta^{i}_{\ell} T^{\ell}_{i}, \tag{A2.4}$$

$$\mathrm{II}_T = \frac{1}{2!}\delta^{ij}_{lm}T^l_i T^m_j ,$$

$$\mathrm{III}_T = \frac{1}{3!}\delta^{ijk}_{lmn}T^l_i T^m_j T^n_k ,$$

and all three invariants are isotropic.

Here we have used the symbols

$$\delta^{ijk}_{lmn} \equiv \varepsilon^{ijk}\varepsilon_{lmn} ,$$

$$\delta^{ij}_{lm} \equiv \delta^{ijn}_{lmn} = \delta^i_l\delta^j_m - \delta^i_m\delta^j_l .$$

The principal directions of a second-order symmetric tensor are the directions determined by the unit vectors $\underset{\sim}{n}$, such that $T^\alpha_\beta n^\beta = Tn^\alpha$, or

(A2.5) $$(T^\alpha_\beta - T\delta^\alpha_\beta)n^\beta = 0 ,$$

and there are three such directions. Since the equations (A2.5) are homogeneous, the nontrivial solutions for $\underset{\sim}{n}$ exist if

(A2.6) $$\det(T^\alpha_\beta - T\delta^\alpha_\beta) = 0 ,$$

which represents a third-order equation in T ,

(A2.7) $$-T^3 + \mathrm{I}_T T^2 - \mathrm{II}_T T + \mathrm{III}_T = 0 ,$$

and the solutions $T_{(\lambda)}$ are the principal values (eigenvalues, proper values) of the tensor $\underset{\sim}{T}$.

If we denote by $\underset{\sim}{n}^{(\alpha)}$ the vectors of a triad reciprocal to the triad of the vectors $\underset{\sim}{n}_{(\alpha)}$ obtained for $\alpha = 1,2,3$ from (A2.5), it is possible to introduce a coordinate transformation so that the new Cartesian coordinates $\bar{z}^{\alpha}$ are colinear with the principal directions,

$$\bar{z}^{\lambda} = n^{(\lambda)}_{\alpha} z^{\alpha} , \qquad (A2.8)$$
$$z^{\alpha} = n^{\alpha}_{(\lambda)} \bar{z}^{\lambda} ,$$

where

$$n^{(\lambda)}_{\alpha} n^{\alpha}_{(\mu)} = \delta^{\lambda}_{\mu} , \quad n^{(\lambda)}_{\alpha} n^{\beta}_{(\lambda)} = \delta^{\beta}_{\alpha} . \qquad (A2.9)$$

The components $\bar{T}^{\lambda}_{\mu}$ of $\underset{\sim}{T}$ with respect to the new coordinates $\bar{z}^{\alpha}$ are

$$\bar{T}^{\lambda}_{\mu} = T^{\alpha}_{\beta} n^{\beta}_{(\mu)} n^{(\lambda)}_{\alpha} ,$$

and according to (A2.5) and (A2.9) we have

$$\bar{T}^{\lambda}_{\mu} = T_{(\mu)} n^{\alpha}_{(\mu)} n^{(\lambda)}_{\alpha} = T_{(\mu)} \delta^{\lambda}_{\mu} . \qquad (A2.10)$$

Hence, the principal values of a tensor $\underset{\sim}{T}$ are its components with respect to a Cartesian coordinate system with coordinate axes colinear with the principal directions. With respect to this system of coordinates the matrix of the tensor $\underset{\sim}{T}$ has only diagonal elements.

The powers of a tensor $\underset{\sim}{T}$ are defined by the expressions

$$\overset{2}{T}{}^{\lambda}_{.\mu} = T^{\lambda}_{.\alpha} T^{\alpha}_{.\mu} ,$$

(A2.11)

$$\overset{3}{T}{}^{\lambda}_{.\mu} = T^{\lambda}_{.\alpha} T^{\alpha}_{.\beta} T^{\beta}_{.\mu} ,$$

$$\cdots\cdots\cdots$$

and from (A2.10) it follows that

$$\overset{2}{\bar{T}}{}^{\lambda}_{\mu} = \overset{2}{T}_{(\mu)} \delta^{\lambda}_{\mu} , \quad \overset{3}{\bar{T}}{}^{\lambda}_{\mu} = \overset{3}{T}_{(\mu)} \delta^{\lambda}_{\mu} , \dots .$$

Since $T(\mu)$ are the solutions of (A2.7) we have obviously

(A2.12) $$\overset{3}{T}_{(\mu)} \delta^{\lambda}_{\mu} = I_T T_{(\mu)} \delta^{\lambda}_{\mu} - II_T T_{(\mu)} \delta^{\lambda}_{\mu} + III_T \delta^{\lambda}_{\mu}$$

or

(A2.13) $$\overset{3}{\underset{\sim}{T}} = I_T \overset{2}{\underset{\sim}{T}} - II_T \underset{\sim}{T} + III_T \underset{\sim}{1} ,$$

which represents the Cayley-Hamilton theorem.

For an antisymmetric tensor $M^{\alpha\beta\gamma} = -M^{\beta\alpha\gamma}$ of the third order, the corresponding second order tensor, according to (A1.32) is given by

(A2.14) $$M_{\lambda}^{\cdot\gamma} = \frac{1}{2} e_{\lambda\alpha\beta} M^{\alpha\beta\gamma} .$$

Because of the nonsymmetry of $\underset{\sim}{M}$ for the construction of the invariants we have to regard besides its components $M_{\lambda}^{\cdot\gamma}$ also the components $M^{\mu}_{\cdot\nu} = g^{\lambda\mu} g_{\mu\nu} M_{\lambda}^{\cdot\gamma}$, which makes the list of invariants larger than the list of invariants of a symmetric second-order tensors. There is one linear invariant,

$$I_M = \delta^k_l M^{\cdot l}_{k} , \qquad (A2.15)$$

but there are two independent quadratic invariants,

$$\begin{aligned} {}^1\mathrm{II}_M &= \frac{1}{2!}\delta^{ij}_{lm} M^{\cdot l}_{i} M^{\cdot m}_{j} , \\ {}^2\mathrm{II}_M &= \frac{1}{2!}\delta^{ij}_{lm} M^{\cdot l}_{i} M^{m}_{\cdot j} , \end{aligned} \qquad (A2.16)$$

and there are eight independent cubic invariants, etc.

If we write for I_M the expression

$$I_M = \frac{1}{2} e_{\alpha\beta\gamma} M^{\alpha\beta\gamma} , \qquad (A2.17)$$

and apply the orthogonal transformation (A2.2) to the components of $\underset{\sim}{M}$, we obtain

$$I_M = \frac{1}{2} e_{\alpha\beta\gamma} Q^{\cdot\alpha}_{\lambda} Q^{\cdot\beta}_{\mu} Q^{\cdot\gamma}_{\nu} \bar{M}^{\lambda\mu\nu} .$$

Since

$$e_{\alpha\beta\gamma} Q^{\cdot\alpha}_{\lambda} Q^{\cdot\beta}_{\mu} Q^{\cdot\gamma}_{\nu} = (\det Q^{\cdot\alpha}_{\lambda}) e_{\lambda\mu\nu} = \pm 1 e_{\lambda\mu\nu} ,$$

and it follows that I_M is a hemitropic invariant.
The invariants ${}^1\mathrm{II}_M$ and ${}^2\mathrm{II}_M$ are isotropic.

The joint invariants of a symmetry tensor $\underset{\sim}{T}$ and of a non-symmetric tensor $\underset{\sim}{M}$ are quadratic

$$\text{(A2.18)} \qquad II_{TM} = T^i_{\ell} M^{\ell}_{.i} = T^i_{\ell} M_i^{.\ell} \; ;$$

cubic

$$\text{(A2.19)} \qquad \begin{aligned} {}^1III_{TM} &= T^i_{\ell} T^{\ell}_m M_i^{.m} \; , \\ {}^2III_{TM} &= T^i_{\ell} M_i^{.m} M_m^{.\ell} \; , \\ {}^3III_{TM} &= T^i_{\ell} M^m_{.i} M_m^{.\ell} \; , \\ {}^4III_{TM} &= T^i_{\ell} M^m_{.i} M^{\ell}_{.m} \; , \\ &\cdots\cdots\cdots \end{aligned}$$

Possible are also other combinations of one symmetric and one non-symmetric second-order tensor, which are not listed in (A2.18,19), but it may easily be verified that the listed invariants (cubic and quadratic) are the only independent invariants. For higher order invariants I have not tried to establish the list of independent invariants.

Among the listed joint invariants, II_{TM} and ${}^1III_{TM}$ are hemitropic, and the remaining invariants are isotropic.

The principal invariants of a symmetric tensor $\underset{\sim}{T}$ may be expressed also in terms of the principal values of $T_{(\lambda)}$,

$$\text{(A2.20)} \qquad \begin{aligned} I_T &= T_{(1)} + T_{(2)} + T_{(3)} \; , \\ II_T &= T_{(2)} T_{(3)} + T_{(3)} T_{(1)} + T_{(1)} T_{(2)} \; , \\ III_T &= T_{(1)} T_{(2)} T_{(3)} \; . \end{aligned}$$

Sometimes it is useful to consider the moments $\overline{II}_T$, $\overline{III}_T$, instead of the principal invariants. The moments are related to the principal invariants by the formulae

$$\begin{aligned} \overline{II}_T &= T^{i}_{.j}T^{j}_{.i} = I_T^2 - 2II_T = \sum_{\alpha=1}^{3} T^2_{(\alpha)} , \\ \overline{III}_T &= T^{i}_{.j}T^{j}_{.k}T^{k}_{.i} = I_T^3 - 3I_T II_T + 3III_T = \sum_{\alpha=1}^{3} T^3_{(\alpha)} . \end{aligned} \tag{A2.21}$$

In the theory of plasticity often is used the so called octaedral invariant ;

$$3\Delta_T = \left[2I_T^2 - 6II_T\right]^{1/2} = \sum_{\alpha>\beta}\left[\left(T_{(\alpha)} - T_{(\beta)}\right)^2\right]^{1/2} . \tag{A2.22}$$

If a tensor is decomposed into its spherical and deviatoric parts,

$$T^{i}_{.j} = \frac{1}{3}I_T\delta^i_j + \left(T^{i}_{.j} - \frac{1}{3}I_T\delta^i_j\right) = {}^{S}T^{i}_{j} + \tau^{i}_{.j} , \tag{A2.23}$$

the principal invariants of the spherical part are

$${}^{S}I_T = I_T, \quad {}^{S}II_T = \frac{1}{3}I_T^2, \quad {}^{S}III_T = \frac{1}{27}I_T^3 \tag{A2.24}$$

and the first invariant of the deviatoric part vanishes identically

$$I_\tau \equiv {}^{D}I_T = 0 . \tag{A2.25}$$

Since (A2.25) represents a relation between nine components of a tensor, it follows that a deviator has only eight independent components.

A second-order tensor can be uniquely decomposed into its symmetryc and antisymmetric parts. For a third order tensor such a decomposition is more involved because we are searching for its irreducible parts. Toupin [462] introduced the following decomposition.

Let M^{ijk} be an arbitrary tensor of the third order. Its irreducible parts are:

the symmetric part

$$\text{(A2.26)} \quad {}_{S}M^{ijk} = M^{(ijk)} = \frac{1}{3!}(M^{ijk} + M^{jki} + M^{kij} + M^{ikj} + M^{jik} + M^{kji}),$$

the antisymmetric part

$$\text{(A2.27)} \quad {}_{A}M^{ijK} = M^{[ijK]} = \frac{1}{3!}(M^{ijk} + M^{jki} + M^{kij} - M^{ikj} - M^{jik} - M^{kji}),$$

the principal parts

$$\text{(A2.28)} \quad \begin{aligned} {}_{P}M^{ijk} &= \frac{1}{3}(M^{ijk} + M^{kji} - M^{jik} - M^{kij}), \\ {}_{\bar{P}}M^{ijk} &= \frac{1}{3}(M^{ijk} + M^{jik} - M^{kji} - M^{jki}). \end{aligned}$$

The symmetric part ${}_{S}\underset{\sim}{M}$ has 10 independent components, the antisymmetric part has 1, and the principal parts ${}_{P}\underset{\sim}{M}$ and ${}_{\bar{P}}\underset{\sim}{M}$ have 8 independent components each, so that the tensor $\underset{\sim}{M}$ is determined by 27 independent components of its irreducible parts, and

$$\underset{\sim}{M} = {}_{S}\underset{\sim}{M} + {}_{A}\underset{\sim}{M} + {}_{P}\underset{\sim}{M} + {}_{\bar{P}}\underset{\sim}{M} \,. \tag{A2.29}$$

A3. Differentiation

If $\underset{\sim}{V}$ is a vector field in E_3 with components V^k and V_ℓ with respect to a coordinate system x^i, the partial derivatives of the vector $\underset{\sim}{V}$ are given by the expressions

$$\frac{\partial \underset{\sim}{V}}{\partial x^m} = \frac{\partial V^k}{\partial x^m}\underset{\sim}{g}_k + V^k \frac{\partial \underset{\sim}{g}_k}{\partial x^m} = \left(\frac{\partial V^k}{\partial x^m} + \begin{Bmatrix} k \\ m\ell \end{Bmatrix} V^\ell\right)\underset{\sim}{g}_k \,, \tag{A3.1}$$

or

$$\frac{\partial \underset{\sim}{V}}{\partial x^m} = \frac{\partial V_\ell}{\partial x^m}\underset{\sim}{g}^\ell + V_\ell \frac{\partial \underset{\sim}{g}^\ell}{\partial x^m} = \left(\frac{\partial V_\ell}{\partial x^m} - \begin{Bmatrix} k \\ \ell m \end{Bmatrix} V_k\right)\underset{\sim}{g}^\ell \,, \tag{A3.2}$$

where

$$V^k_{\ ,m} \equiv \frac{\partial V^k}{\partial x^m} + \begin{Bmatrix} k \\ m\ell \end{Bmatrix} V^\ell \,, \tag{A3.3}$$

$$V_{\ell,m} \equiv \frac{\partial V_\ell}{\partial x^m} - \begin{Bmatrix} k \\ m\ell \end{Bmatrix} V_k \,, \tag{A3.4}$$

represent the <u>covariant derivatives</u> of co- and contravariant components of the vector field $\underset{\sim}{V}$.

The quantities

(A3.5) $$[\ell m,n] \equiv \frac{\partial \underset{\sim}{g}_m}{\partial x^\ell}\underset{\sim}{g}_n = \frac{1}{2}\left(\frac{\partial g_{mn}}{\partial x^\ell} + \frac{\partial g_{n\ell}}{\partial x^m} - \frac{\partial g_{\ell m}}{\partial x^n}\right)$$

are the Christoffel symbols of the first kind, and

(A3.6) $$\left\{ {k \atop m\ell} \right\} \equiv g^{kn}[\ell m,n] = \frac{\partial \underset{\sim}{g}_\ell}{\partial x^m}\underset{\sim}{g}^k$$

are the Christoffel symbols of the second kind.

In general, if $\underset{\sim}{T}$ is a tensor of contravariant order p and covariant order q, the covariant derivatives of its components are tensors of contravariant order p and covariant order $q+1$,

$$T^{i_1\ldots i_p}_{\ j_1\ldots j_q,k} = \frac{\partial}{\partial x^k} T^{i_1\ldots i_p}_{\ j_1\ldots j_q} +$$

(A3.7) $$+ \sum_{\alpha=1}^{p} \left\{ {i_\alpha \atop k\,\ell} \right\} T^{i_1\ldots i_{\alpha-1}\,\ell\, i_{\alpha+1}\ldots i_p}_{\ j_1\ldots j_q} -$$

$$- \sum_{\beta=1}^{q} \left\{ {\ell \atop k\,j_\beta} \right\} T^{i_1\ldots i_p}_{\ j_1\ldots j_{\beta-1}\,\ell\, j_{\beta+1}\ldots j_q} .$$

For the sake of brevity we write sometimes for partial derivatives

(A3.8) $$\frac{\partial}{\partial x^m} = \partial_m .$$

The covariant differential of a tensor $\underset{\sim}{T}$ is a tensor of the same order, defined by the expression

$$\delta T^{\cdots}_{\cdots} = T^{\cdots}_{\cdots,k} dx^k \,. \tag{A3.9}$$

Let $\underset{\sim}{T}$ be a time-independent tensor field. The absolute time derivatives of the components of $\underset{\sim}{T} = \underset{\sim}{T}(\underset{\sim}{x},t)$ are defined by the formula

$$\frac{DT^{\cdots}_{\cdots}}{dt} = \frac{\partial T^{\cdots}_{\cdots}}{\partial t} + T^{\cdots}_{\cdots,k}\frac{dx^k}{dt} = \dot{T}^{\cdots}_{\cdots} \,. \tag{A3.10}$$

For double tensor fields we define partial and total covariant derivatives. If $T^k_{.K}(\underset{\sim}{x},\underset{\sim}{X})$ is such a tensor, the partial covariant derivatives are defined by

$$T^k_{.K,\ell} = \frac{\partial T^k_{.K}}{\partial x^\ell} + \left\{ {k \atop \ell m} \right\} T^m_{.K} \,, \tag{A3.11}$$

$$T^k_{.K,L} = \frac{\partial T^k_{.K}}{\partial X^L} - \left\{ {M \atop LK} \right\} T^k_{.M} \,. \tag{A3.12}$$

If there is a mapping $\underset{\sim}{x} = \underset{\sim}{x}(\underset{\sim}{X})$, the <u>total covariant derivatives</u> with respect to x^ℓ and X^L are defined as a generalization of the classical rule

$$T^k_{.K;\ell} = T^k_{.K,\ell} + T^k_{.K,L} X^L_{;\ell} \,, \tag{A3.13}$$

$$T^k_{.K;L} = T^k_{.K,L} + T^k_{.K,\ell}\, x^\ell_{;L} \,, \tag{A3.14}$$

where $X^L_{;\ell}$ and $x^\ell_{;L}$ are the gradients of the mapping $\underset{\sim}{x} \leftrightarrow \underset{\sim}{X}$. The chain rule of ordinary differential calculus also holds for total covariant differentiation,

(A3.15)
$$T^{\cdots}_{\cdots;K} = T^{\cdots}_{\cdots;k} x^k_{;K}$$
$$T^{\cdots}_{\cdots;k} = T^{\cdots}_{\cdots;K} X^K_{;k} .$$

A4. Linearly Connected Spaces

Let V^α be components of a vector field in E_3, referred to a system of Cartesian coordinates and let us perform a parallel displacement of the vector $\underset{\sim}{V}$ from a point $\underset{\sim}{z}$ to a neighbouring point $\underset{\sim}{z} + d\underset{\sim}{z}$. The components of the vector $\underset{\sim}{V}$ will remain unchanged. Denoting by $\overset{*}{d}V^\alpha$ the change of the components at a parallel displacement along $d\underset{\sim}{z}$ we may write

(A4.1)
$$\overset{*}{d}V^\alpha = 0 .$$

However, when the vector field V^α is referred to an arbitrary system of curvilinear coordinates x^i, (A4.1) will obtain the form

(A4.2)
$$\overset{*}{d}V^k = -\left\{ {k \atop \ell m} \right\} V^m dx^\ell .$$

The vector field V^k at a point $\underset{\sim}{x} + d\underset{\sim}{x}$ has the components

$$V^k(\underset{\sim}{x} + d\underset{\sim}{x}) = V^k(\underset{\sim}{x}) + \partial_\ell V^k dx^\ell + \ldots . \qquad (A4.3)$$

The difference between the field value of the vector $\underset{\sim}{V}$ at $\underset{\sim}{x} + d\underset{\sim}{x}$ and $V^k + dV^k$ is the covariant differential,

$$\delta V^k = V^k(\underset{\sim}{x} + d\underset{\sim}{x}) - \overset{*}{V}{}^k = \left(\partial_\ell V^k + \left\{ {k \atop \ell m} \right\} V^m\right) dx^\ell . \qquad (A4.4)$$

According to (A4.2) parallelism in Euclidean space is defined (in the sense of differential geometry) as a linear connection of the increment $\overset{*}{d}V^k$ of the components of the vector V^k and the components dx^ℓ of the displacement.

The law (A4.2) may be generalized writing

$$dV^k = -\Gamma^k_{\ell m} V^m dx^\ell , \qquad (A4.5)$$

where $\Gamma^k_{\ell m}$ are arbitrary functions of position and are called <u>coefficients of connection</u> of a <u>linearly connected</u> space L_3.

In general, the coefficients $\Gamma^k_{\ell m}$ are not symmetric, and the antisymmetric part $S_{\ell m}^{\cdot\cdot k} = \Gamma^k_{[\ell m]}$ is the <u>torsion tensor</u> of the space L_3.

Generalizing the rules for covariant differentiation to linearly connected spaces we may write for the covariant derivatives of a contravariant vector

$$V^k_{,\ell} = \partial_\ell V^k + \Gamma^k_{\ell m} V^m , \qquad (A4.6)$$

and from the requirement that $V^k_{,\ell}$ transforms like a mixed second-order tensor we obtain the transformation law for the coefficients of connection:

$$\bar{\Gamma}^i_{jk} = \Gamma^\ell_{mn} \frac{\partial x^m}{\partial \bar{x}^j} \frac{\partial x^n}{\partial \bar{x}^k} \frac{\partial \bar{x}^i}{\partial x^\ell} + \frac{\partial \bar{x}^i}{\partial x^\ell} \frac{\partial^2 x^\ell}{\partial \bar{x}^j \partial \bar{x}^k} =$$

(A4.7)

$$= \Gamma^\ell_{mn} \frac{\partial x^m}{\partial \bar{x}^j} \frac{\partial x^n}{\partial \bar{x}^k} \frac{\partial \bar{x}^i}{\partial x^\ell} - \frac{\partial x^m}{\partial \bar{x}^j} \frac{\partial x^n}{\partial \bar{x}^k} \frac{\partial^2 \bar{x}^i}{\partial x^m \partial x^n} .$$

From (A4.7) it follows that $S_{mn}^{\cdot\cdot\ell}$ is a tensor indeed.

Parallelism in an L_n is, according to (A4.5), defined only for infinitesimal displacements. If ABC is a curve in L_3, the total increment ΔV^k of the components V^k of a vector transported parallely from A to C along the curve will be

$$\Delta V^k = \int_{ABC} \overset{*}{d} V^k = -\int_{ABC} \Gamma^k_{\ell m} V^m dx^\ell .$$

If $AB'C$ is another curve connecting the points A and C, the increment of the components of the vector V^k along this curve will be

$$\Delta'' V^k = \int_{AB'C} \overset{*}{d} V^k ,$$

and the increments $\Delta' V^k$ and $\Delta'' V^k$ are, in general, not equal, i.e. the integral along the closed contour $ABCB'A$ is not vanishing,

$$\Delta V^k = \oint_{ABCB'A} \overset{*}{d}V^k = -\oint \Gamma^k_{\ell m} V^m dx^\ell = \Delta' V^k - \Delta'' V^k .$$

Denoting $-\Gamma^k_{\ell m} V^m$ by f^k_ℓ and applying the Stokes theorem,

$$\oint f^k_\ell dx^\ell = \iint_F f^k_{[\ell,m]} dF^{m\ell} \tag{A4.8}$$

where F is the surface enclosed by the contour $ABCB'A$ and $dF^{m\ell}$ are components of the surface element, $\Delta F^{m\ell} = -\Delta F^{\ell m}$, we have

$$\Delta V^k = \iint_F R_{nm\ell}^{\cdot\cdot\cdot k} V^\ell dF^{mn} , \tag{A4.9}$$

where

$$R_{nm\ell}^{\cdot\cdot\cdot k} = \partial_n \Gamma^k_{m\ell} - \partial_m \Gamma^k_{n\ell} + \Gamma^k_{nt}\Gamma^t_{m\ell} - \Gamma^k_{mt}\Gamma^t_{n\ell} \tag{A4.10}$$

is the Riemann-Christoffel curvature tensor.

If $R_{nm\ell}^{\cdot\cdot\cdot k}$ vanishes at all points of the space, we say that this space is with absolute parallelism (or with teleparallelism).

In Euclidan spaces the fundamental tensor g_{ij} is covariant constant, i.e. its covariant derivatives are identically equal to zero. If an L_3 admits a symmetric covariant constant vector field g_{ij} , we say that the space L_3 is metric. Let us assume that an L_3 with the coefficients of connection $\Gamma^k_{\ell m}$ is metric and that its fundamental metric tensor is g_{ij} , then we have

$$g_{ij,k} = \partial_k g_{ij} - \Gamma^l_{ki} g_{li} - \Gamma^l_{kj} g_{il} = 0 . \tag{A4.11}$$

The integrability conditions of (A4.11) are

$$(\partial_l \partial_k - \partial_k \partial_l) g_{ij} = 0 ,$$

and after some calculations they reduce to

$$R^{\cdot\cdot\cdot\cdot}_{nm(lk)} = 0 . \tag{A4.12}$$

Hence, if the Riemann-Christoffel tensor for a linear connection Γ^k_{lm} is symmetric with respect to the second pair of indices, the connection is metric.

The linearly connected space is Euclidean if:

1° the coefficients of connection are symmetric,

2° it is a metric space,

3° the fundamental form of the space

$$ds^2 = g_{ij} dx^i dx^j \tag{A4.13}$$

is positive definite, and

4° if the Riemann-Christoffel tensor vanishes everywhere in the space. If all these conditions are satisfied, it is possible to find a coordinate transformation

$$\begin{aligned} x^i &= x^i(z^1, z^2, z^3) \\ z^\alpha &= z^\alpha(x^1, x^2, x^3) , \end{aligned} \tag{A4.14}$$

such that the fundamental tensor with respect to the new coordinate system z^α obtains the form

$$g_{\alpha\beta} = \frac{\partial x^i}{\partial z^\alpha}\frac{\partial x^j}{\partial z^\beta}g_{ij} = \delta_{\alpha\beta} . \tag{A4.15}$$

In some problems we have to deal with the correspondence of a set of points of Euclidean space with a set of points of a linearly connected space L_3 . If x^α is a system of coordinates in Euclidean space, and u^α a system of coordinates in L_3 there do not exist 1:1 finite mappings of the form

$$\begin{aligned} x^i &= x^i(u^1,u^2,u^3) , \\ u^\alpha &= u^\alpha(x^1,x^2,x^3) , \end{aligned} \tag{A4.16}$$

but only the local mappings of infinitesimal elements dx^i and du^α,

$$dx^i = \Phi^i_{(\alpha)}du^\alpha . \tag{A4.17}$$

We assume that the relations (A4.17) are linearly independent,

$$\det \Phi^i_{(\alpha)} \neq 0 \tag{A4.18}$$

so that there exist the inverse relations

$$du^\alpha = \Phi^{(\alpha)}_i dx^i . \tag{A4.19}$$

The integrability conditions of (A4.17)

(A4.20) $$2S^{(\alpha)}_{ij} = \partial_j \Phi^{(\alpha)}_i - \partial_i \Phi^{(\alpha)}_j = 0$$

and those conditions are, in general, not satisfied.

The vectors $\underset{\sim}{\Phi}_{(\alpha)}$ constitute in E_3 three vector fields and at each point there are lines the tangents of which are colinear with the vectors $\underset{\sim}{\Phi}_{(\alpha)}$. The differential equations of these lines are

(A4.21) $$\frac{dx^1}{\Phi^1_{(\alpha)}} = \frac{dx^2}{\Phi^2_{(\alpha)}} = \frac{dx^3}{\Phi^3_{(\alpha)}} .$$

Let us assume that there is a linearly connected space with the coefficients of connection Γ^k_{ij} such that the vector fields $\Phi^i_{(\alpha)}$ constitute fields of absolutely parallel vectors, i.e. with respect to the connection considered, the vectors $\underset{\sim}{\Phi}_{(\alpha)}$ are covariant constant everywhere in the space,

(A4.22) $$\partial_m \Phi^k_{(\alpha)} + \Gamma^k_{m\ell} \Phi^\ell_{(\alpha)} = 0 .$$

Transvection of this with $\Phi^{(\alpha)}_n$ and using the relations

(A4.23) $$\Phi^{(\alpha)}_n \Phi^k_{(\alpha)} = \delta^k_n , \quad \Phi^{(\alpha)}_\ell \Phi^\ell_{(\beta)} = \delta^\alpha_\beta$$

we obtain

(A4.24) $$\Gamma^k_{mn} = -\Phi^{(\alpha)}_n \partial_m \Phi^k_{(\alpha)} = \Phi^k_{(\alpha)} \partial_m \Phi^{(\alpha)}_n .$$

It may easily be verified that substituting Γ_{mn}^{k} from (A2.24) into the expression (A4.10) for the components of $R_{nml}^{\cdot\cdot\cdot k}$ it will identically vanish. According to (A4.12) it follows that the conditions for the space considered to be metric are identically fulfilled.

From the preceding it follows that it is always possible to associate a linearly connected metric space to a non- integrable mapping, and the torsion of this space does not necessarily vanish.

The torsion tensor of the connection (A4.23) is given by

$$S_{mn}^{\cdot\cdot k} = \Phi_{(\alpha)}^{k}\partial_{[m}\Phi_{n]}^{(\alpha)} = \Phi_{(\alpha)}^{k} S_{mn}^{(\alpha)} , \qquad (A4.25)$$

and it is obvious that the space associated to a non-integrable mapping will be Euclidean only if the torsion vanishes i.e. if the mapping is integrable (this is a necessary, but not a sufficient condition).

The quantities obtained by transvecting vectors, tensors etc. of Euclidean space with the components of the vectors $\underset{\sim}{\Phi}_{(\alpha)}$, $\underset{\sim}{\Phi}^{(\alpha)}$, e.g.

$$V^{(\alpha)} = V^{i}\Phi_{i}^{(\alpha)} , \quad T_{\cdot(\beta)}^{(\alpha)} = T_{\cdot j}^{i}\Phi_{i}^{(\alpha)}\Phi_{(\beta)}^{j} \qquad (A4.26)$$

are often called <u>non-holonomic components</u> of those quantities.

A5. Modified Divergence Theorem for Incompatible Deformations Variations.

Since there are no integrable mappings of a non--Riemannian space upon the Euclidean space, a straightforward application of the divergence theorem to the integrals of the form

$$J = \oint_S T_i^{\cdot j} v^i ds_j \tag{A5.1}$$

is impossible. We assume that $\underset{\sim}{T}$ is any regular differentiable tensor field in E_3. S is the surface bounding an arbitrary volume v of a body B.

The whole region v may be divided into a number of small elements Δv_α with bounding surfaces ΔS_α and we have

$$J = \sum_{\alpha=1}^{n} J_\alpha = \sum_{\alpha=1}^{n} \oint_{\Delta S_\alpha} T_i^{\cdot j} v^i ds_j . \tag{A5.2}$$

For Cartesian coordinates x^i we may choose Δv_α to be cuboids with edges parallel to the Cartesian axes x^1, x^2, x^3 such that the sides of the cuboids are $\Delta x^1, \Delta x^2, \Delta x^3$. If we put $T_i^{\cdot j} v^i = T^j$, we have

$$J_\alpha = \int_{\Delta S_\alpha^1} T^1 ds_1 + \int_{\Delta S_\alpha^2} T^2 ds_2 + \int_{\Delta S_\alpha^3} T^3 ds_3 + \tag{A5.3a}$$

$$+ \int_{-\Delta S^1_\alpha} T^1 ds_1 + \int_{-\Delta S^2_\alpha} T^2 ds_2 + \int_{-\Delta S^3_\alpha} T^3 ds_3 . \qquad \text{(A5.3b)}$$

Putting $x = x^1$, $y = x^2$; $z = x^3$, the faces Δs^1_α, Δs^2_α, Δs^3_α will be orthogonal to the axes x^1, x^2, x^3. Thus on the faces Δs^1_α we have

$$\Delta s^1_\alpha : \qquad T^1 = T^1(x + \Delta x, y, z) ,$$

and on the face

$$-\Delta s^1_\alpha : \qquad T^1 = T^1(x, y, z) .$$

Similarly

$$\Delta s^2_\alpha : \qquad T^2 = T^2(x, y + \Delta y, z) ,$$

$$-\Delta s^2_\alpha : \qquad T^2 = T^2(x, y, z) ,$$

$$\Delta s^3_\alpha : \qquad T^3 = T^3(x, y, z + \Delta z) ,$$

$$-\Delta s^3_\alpha : \qquad T^3 = T^3(x, y, z) .$$

Hence, for the pair of integrals

$$J^1_\alpha = \int_{\Delta S^1_\alpha} T^1 ds_1 - \int_{-\Delta S^1_\alpha} T^1 ds_1 = \int_{\Delta S^1_\alpha} \Delta T^1 ds_1 \qquad \text{(A5.4)}$$

we have

$$\text{(A5.5)} \qquad J^1_\alpha = \int_y^{y+\Delta y} \int_z^{z+\Delta z} \left[T^1(x+\Delta x,y,z) - T^1(x,y,z)\right] dy\,dz \,.$$

However, for a regular tensor field $T_i^{\cdot j}$ and for sufficiently small Δx^i we have

$$\text{(A5.6)} \qquad T^1_i(x+\Delta x,y,z) = T^1_i(x,y,z) + \partial_1 T^1_i \Delta x + \dots \,.$$

For the velocities v^i we have (4.1.16)

$$\text{(A5.7)} \qquad v^i\big|_{x+\Delta x,y,z} = v^i\big|_{x,y,z} + \Delta v^i = v^i + \dot{\Phi}^i_{(\lambda)} \Phi^{(\lambda)}_1 \Delta x^1 \,.$$

The difference ΔT^1 in (A5.6) obtains now the form

$$\text{(A5.8)} \qquad \Delta T^1 = (T^1_i \dot{\Phi}^i_{(\lambda)} \Phi^{(\lambda)}_i + \partial_1 T^{\cdot 1}_i v^i) \Delta x^1 \,.$$

For infinitesimal elements Δv_α the mean-value theorem may be applied to the integrals J^1_α , and it gives

$$J^1_\alpha = (T^{\cdot 1}_i \dot{\Phi}^i_{(\lambda)} \Phi^{(\lambda)}_1 + \partial_1 T^{\cdot 1}_i v^i) \Delta x \Delta y \Delta z \,,$$

and, in general,

$$J_\alpha = (T^{\cdot j}_i \dot{\Phi}^i_{(\lambda)} \Phi^{(\lambda)}_j + \partial_j T^{\cdot j}_i v^i) dv \,.$$

When $\Delta v_\alpha \to 0$ and $n \to \infty$ the sum (A5.2) becomes the volume integral over v and for any curvilinear system of coordinates we may finally write

$$J = \oint_{S} T_{i}^{\cdot j} v^{i} ds_{j} = \int_{v} (\nabla_{j} T_{i}^{\cdot j} + T_{i}^{\cdot j} \dot{\Phi}^{i}_{(\lambda)} \Phi^{(\lambda)}_{j}) dv . \qquad \text{(A5.9)}$$

When we deal with the variations δx^{i} of coordinates (cf. Stojanović [421]), it follows from (4.1.12) that

$$\Delta \delta x^{i} = \delta x_{2}^{i} - \delta x_{1}^{i} = \delta \Phi^{i}_{(\lambda)} \Delta u^{\lambda} = \delta \Phi^{i}_{(\lambda)} \Phi^{(\lambda)}_{j} \Delta x^{j} .$$

When the directors $\overset{(\lambda)}{d}_{i}$ are compared at two points, say P and Q of a body, we have for sufficiently near to one another points

$$\overset{(\lambda)}{d}_{i}\{Q\} - \overset{(\lambda)}{d}_{i}\{P\} = \Delta \overset{(\lambda)}{d}_{i} = \overset{(\lambda)}{d}_{i,j} \Delta x^{j} .$$

The variation of this difference will be

$$\delta \overset{(\lambda)}{d}_{i}\{Q\} - \delta \overset{(\lambda)}{d}_{i}\{P\} = \delta \Delta \overset{(\lambda)}{d}_{i} = \Delta x^{j} \delta \overset{(\lambda)}{d}_{i,j} + \overset{(\lambda)}{d}_{i,j} \delta \Delta x^{j} .$$

But we also have

$$\delta \Delta \overset{(\lambda)}{d}_{i} = \delta \overset{(\lambda)}{d}_{i}\{Q\} - \delta \overset{(\lambda)}{d}_{i}\{P\} = (\delta \overset{(\lambda)}{d}_{i})_{,j} \Delta x^{j} ,$$

and

$$(\delta \overset{(\lambda)}{d}_{i})_{,j} \Delta x^{j} = \delta (\overset{(\lambda)}{d}_{i,j}) \Delta x^{j} + \overset{(\lambda)}{d}_{i,\ell} \Phi^{(\mu)}_{j} \delta \Phi^{\ell}_{(\mu)} \Delta x^{j} .$$

Since this expression must be valid for arbitrary Δx^{j}, we finally have

$$(\delta \overset{(\lambda)}{d}_{i})_{,j} = \delta (\overset{(\lambda)}{d}_{i,j}) + \overset{(\lambda)}{d}_{i,\ell} \Phi^{(\mu)}_{j} \delta \Phi^{\lambda}_{(\mu)} . \qquad \text{(A5.10)}$$

References

Besides the papers quoted in the text, this list of references contains also references to other work dealing with polar continua. The desire was to make as complete as possible a bibliography on mechanics of polar continua, and here are listed all papers and books treating this matter according to the knowledge of the author. Unfortunately, there is a number of important contributions of whose existence I was not aware at the moment when the list was completed.

For the majority of journals usual abbreviations are used. E.g. PMM = Prikladnaja Matematika i Mekhanika, Int. J. Engng. Sci. = International Journal of Engineering Sciences, Arch. Rat. Mech. Anal. = Archives for Rational Mechanics and Analysis, App. Math.Mech. = Applied Mathematics and Mechanics (English translation of the Soviet Journal PMM), etc.

[1] J.D. Achenbach: Free vibrations of a layer of micropolar continuum. Int.J. Engng. Sci. 7, 1025-1040 (1969)

[2] J.D. Achenbach, G. Herrmann: Wave motion in solids with lamellar structuring. Dynamics of Structured Solids, pp. 23-46 Publ. by ASME .

[3] J.D. Achenbach, G. Herrmann: Effective stiffness for a laminated composite. Developments in Mechanics. Proc. 10th Midwestern Mechanics Conference, Colorado (1968).

[4] E.L. Aero, A.N. Bul'gin, E.V. Kuvshinskii: Asimmetricheskaja gidromekhanika. Prikl. Mat. Mekh. 29, 297-308 (1964).

[5] E.L. Aero, E.V. Kuvshinskii: Osnovnije Uravneneija Teorii Uprugosti Sred S Vrashchtal'nim Vzaimodeistviem Chastic. Fiz. tv. tela 2, 1399-1409 (1960).

[6] E.L. Aero, E.V. Kuvshinskii: Kontinualnaja Teorija Asimmetricheskoi Uprugosti. Ravnovesije Izotropnogo Tela. Fiz. tv. tela 6, 2689-2699 (1964).

[7] E.F. Afanas'ev, V.I. Nikolaevskii: K postroeniju asimmetricnoi gidrodinamiki suspenzii s vrascajuimisja tverdimi casticami. Problemi Gidrodinamiki i mehaniki splosnoi sred'i, Sedov aniversary volume, Moskwa 1969, Izdateljstvo "Nauka", pp. 17-24.

[8] H.R. Aggarwal, R.C. Alverson: The effect of couple-stresses on the diffraction of plane elastic waves by cylindrical discontinuities. Int. J. Solids. Structure 5, 491-511 (1969).

[9] J.B. Alblas: Continuum mechanics of media with internal structure. Simposia Mathematica 1, 229-251 (1969).

[10] S.J. Allen, C.N. DeSilva: A theory of transversally isotropic fluids. J. Fluid Mech. 24, 801 (1966).

[11] S.J. Allen, C.N. DeSilva: Theory of simple deformable directed fluids. Phys. Fluids 10, 2551-2555 (1967).

[12] S.J. Allen, K.A. Kline: A Theory of mixtures with microstructure. ZAMP 20, 145-155 (1969).

[13] K. Anthony, U. Essmann, A. Seeger, H. Träuble: Disclinations and the Cosserat continuum with incompatible rotations. Mechanics of Generalized Continua. Proc. IUTAM Symposium Freudenstadt-Stuttgart (1967). pp. 355-358.

[14] T. Ariman: On the stresses around a circular hole in micropolar elasticity. Acta Mech. 4, 216-229 (1967).

[15] T. Ariman: Micropolar and dipolar fluids. Int. J. Engng. Sci. 6, 1-8 (1968).

[16] T. Ariman, A.S. Cakmak: Couple stresses in Fluids. Phys. Fluids 10, 2497-2499 (1967).

[17] T. Ariman, A.S. Cakmak, L.R. Hill: Flow of micropolar fluids between two concentric cylinders. Phys. Fluids 10, 2545-2550 (1967).

[18] T. Ariman, A.S. Cakmak: Some basic flows in micropolar fluids. Rheologica Acta 7, 236-242 (1968).

[19] A. Askar, A.S. Cakmak: A structural model of micropolar continuum. Int. J. Engng. Sci. 6, 583-590 (1968).

[20] A.Askar, A.S. Cakmak, T. Ariman: Linear theory of hereditary micropolar materials. Int. J. Engng. Sci. 6, 283 (1968).

[21] D.V. Babich: Postankova dinamicheskoi zadachi dlja obolochek v prostranstve Cosserat. Dinamika sistem tverd. tel. Trudi seminara po dinamike In-ta mekhanike AN USSR 1965. pp. 3-8. Kiev, (1966).

[22] D.V. Babich: Nekatorie dinamicheskie zadachi teorii plastin s uchetom momentnih naprjazhenii. Dinamika sistem tverd. i. zhidk. tel. Trudi seminara po dinamike In-ta mekhanike AN USSR 1965. pp. 9-15 Kiev, (1966).

[23] D.V. Babich: Basic equations of motion of a shell taking into account asymmetry of the stress tensor. Priklandnaja Mekhanika, 2, 41-48 (1966).

[24] D.V. Babich: The effect of couple-stresses on the free vibrations of a cylindrical shell. Priklandnaja Mekhanika, 3, 39-44 (1967).

[25] M.M. Balaban, A.E. Green, P.M. Naghdi: J. Math. Phys. 8, 1026 (1967).

[26] C.B. Banks: Several problems demonstrating the effects of couple-stresses and a lattice analogy. Ph. D. Thesis, Univ. of Kansas, July 1966.

[27] C.B. Banks, M. Sokolowski: On certain two-dimensional applications of the couple-stress theory. Int. J. Solids. Structures 4, 15-29 (1968).

[28] M.F. Beatty: A reciprocal theorem in the linearized theory of couple-stresses. Acta Mechanica 3, 154-166 (1967).

[29] S.I. Ben-Abraham: Generalized stress and Nonriemannian geometry. Reprint: Fundamental aspects of dislocation theory - NBS Special Publication 317, pp. 943-962, Washington 1970.

[30] Ch. W. Bert: Influence of couple stresses on stress concentration. Experimental Mechanics, 3, 307-308 (1963).

[31] C.W. Bert, F.J. Appl.: General solution for two-dimensional couple-stress elasticity. AIAA journal 6, 968-969 (Technical Notes), (1968).

[32] M.A. Biot: Rheological stability with couple stresses and its application to geological folding. Proc. Roy. Soc. London (A) 298, 1455, 402-423 (1967).

[33] D. Blagojević: On general solutions in the theory of elastic materials of grade two. (In Serbocroatian) Proc. IX Yugoslav Congress of Mechanics, Split 1968. pp. 37-46 Belgrade 1969.

[34] D. Blagojević: On exact solutions in the theory of polar elastic materials. (In Serbocroatian). Doct. Diss. Univ. of Belgrade (1969).

[35] D. Blagojević, R. Stojanović: A contribution to the problem of constitutive relations for elastic materials of grade two. (In Serbocroatian). Tehnika (Sect. Civil Engineering) 2, 240-244 (1969).

[36] J.L. Bleustein: Effects of micro-structure on the stress concentration at a spherical cavity. Int. J. Solids Structures 2, 83-104 (1966).

[37] J.L. Bleustein: A note on the boundary conditions of Toupin's strain-gradient theory. Int. J. Solids Structures 3, 1053-1057 (1967).

[38] J.L. Bleustein, A.E. Green: Dipolar Fluids. Int. J. Engng. Sci. 5, 323-340 (1967).

[39] S. Bodaszewski: On the asymmetric state of stress and its applications to the mechanics of continuous media. (In Polish). Arch. Mech. Stosowanej 5, 351-392 (1953).

[40] D.B. Bogy, E. Sternberg: The effect of couple-stress on singularities due to discontinuous loadings. Int. J. Solids Structures 3, 757-770 (1967).

[41] D.B. Bogy, E. Sternberg: The effect of couple-stress on the corner singularity due to an asymmetric shear loading. International Journal of Solids and Structures, 4, 159-172 (1968).

[42] E.H. Brandt: Leerlinie und Stapefehler im Flusslinien-gitter der Supraleiter 2. Art. in der London-Nähe rung. (Dissertation, Teil II) Phys. Stat. Sol. 36, 371-379 (1969).

[43] E.H. Brandt: Die Behandlung von Fehlern im Flusslinien-gitter der Supraleiter 2. Art. in der Nähe von H_{c2} (Dissertation, Teil III) Phys. Stat. Sol. 36, 381-391 (1969).

[44] E.H. Brandt: Die Leerlinie im Flussliniengitter der Supra leiter 2. Art in der Nähe von H_{c2}. (Dissertation, Teil IV) Phys. Stat. Sol. 36, 393-406 (1969).

[45] A. Bressan: Qualche teorema di cinematica delle deforma-zioni finite. Atti. Ist. Veneto Sci. Lett. Art., Clas. di sci. mat. nat. 121, 235-244 (1962/3).

[46] A. Bressan Sui sistemi continui nel caso asimmetrico. I° Ann. Mat. pura ed applic. 61, 169-190 (1963).

[47] A. Bressan Sui sistemi continui nal caso asimmetrico. II° Ann. di Matem. s. IV 62, 190-222 (1963).

[48] A. Bressan: Coppie di contatto in relatività. I. Ann. Sc. Norm. Sup. Pisa 20, 63-99 (1966).

[49] A. Bressan: Elasticità relativistica con coppie di contat to. Ricerche di Matematica 15, 169-190 (1966).

[50] A. Bressan: Sistemi polari in relatività. Symposia Mathe-matica 1, 289-311 (1968).

[51] A. Bressan: On the influence of gravity on elasticity. Meccanica 4, 195-206 (1969).

[52] A.N. Bulygin and E.V. Kuvshinskii: Plane strain in the asymmetric theory of elasticity. (In Russian) Prikl. Mat. Mekh. 31, 569 (1967).

[53] D.E. Carlson: Stress functions for plane problems with couple-stresses. ZAMP 17, 789-792 (1966).

[54] D.E. Carlson: Stress-functions for couple and dipolar stresses. Quaterly of Appl. Math. 24, 29-35 (1966)

[55] D.E. Carlson: On Günther's stress-function for couple-stresses. Quart. Appl. Math. 25, 139 (1967).

[56] D.E. Carlson: On General solution of equilibrium of a Cosserat continuum. J. Appl. Mech. (Trans. ASME, ser. E.) 34, 245-246 (1967).

[57] M.M. Carroll: Plane waves in non-simple elastic solids, Int. Journ. Solids Structures 5, 109-116 (1969).

[58] M.M. Carroll: Controllable motions of incompressible non--simple materials. Arch. rat. Mech. Anal. 34, 128--142 (1969).

[59] S.K. Chakrabarti, W.L. Wainwright: On the forumaltion of constitutive relations. Int. J. Engng. Sci. 7, 601-613 (1969).

[60] R.S. Chauhan: Couple-stresses in a curved bar. Int. J. Engng. Sci 7, 895-903 (1969).

[61] P.J. Chen, M.E. Gurtin, W.O. Williams: On the thermodynamics of non-simple elastic materials with two temperatures. ZAMP 20, 107-112 (1969).

[62] W.D. Claus, Jr. A.C. Eringen: On dislocations, plasticity and micromorphic mechanics. Techn. Report N° 10, Princeton Univ., Dept. of Aerospace and Mechanical Sciences. October 1969. pp. 1-69.

[63] H. Cohen: A nonlinear theory of elastic directed curves. Int. J. Engng. Sci., 4, 511-524, (1966).

[64] H. Cohen, C.N. DeSilva: Nonlinear theory of elastic surfaces. J. Math. Phys. 7, 246-253 (1966).

[65] H. Cohen, C.N. DeSilva: Nonlinear theory of elastic directed surfaces. J. Math. Phys. 7, 960-966 (1969).

[66] H. Cohen, C.N. DeSilva: A direct nonlinear theory of elastic membranes. Meccanica 3, 141-145 (1969).

[67] B.D. Coleman: Thermodynamics of materials with memory. Arch. Rat. Mech. Anal. 17, 1-46 (1964).

[68] B.D. Coleman: Simple liquid crystals. Arch. Rat. Mech. Anal. 20, 41 (1965).

[69] D.W. Condiff, J.S. Dahler: Fluid mechanical aspects of antisymmetric stress. Phys. Fluids 7, 842 (1964).

[70] T.S. Cook, Y. Weitsman: Strain-gradient effects around spherical inclusions and cavities. Int. J. Solids Structures 2, 393-406 (1966).

[71] E. et F. Cosserat: Sur la mécanique générale. C.R. Acad. Sci. Paris 145, 1139-1142 (1907).

[72] E. et F. Cosserat: Sur la théorie des corps minces. C. R. Acad. Sci. Paris, 146, 169-172 (1908).

[73] E. et F. Cosserat: La théorie des corps deformables. Paris, (1909).

[74] S.C. Cowin: Mechanics of Cosserat Continua. Doc. Diss. Pennsylvania State Univ., (1962), 158 pp.

[75] S.C. Cowin: Stress functions for Cosserat elasticity. Int. J. Solids Structures 6, 389-409 (1970).

[76] M.J. Crochet: Compatibility equations for a Cosserat surface. Journal de Mécanique 6, 593-600, (1967).

[77] M.J. Crochet, P.M. Naghdi: Large deformation solutions for an elastic Cosserat surface. Int. J. Engng. Sci. 7, 309-335, (1969).

[78] J.S. Dahler, L.E. Scriven: Theory of structured continua. I. General consideration of angular momentum and polarization. Proc. Roy. Soc. A 275, 504-527 (1963).

[79] D.A. Danielson: Simplified intrinsic equations for arbitrary elastic shells. Int. J. Engng. Sci. 8, 251--259 (1970).

[80] F.D. Day, Y. Weitsman: Strain gradient effects in microlayers. J. Engng. Mech. Division, Proc. ASCE 92, 67-86 (1966).

[81] N. Van D'ep: Equations of a fluid boundary layer with couple stresses (in Russian) Prikl. Mat. Mekh. 32, 748-753 (1968).

[82] H. Deresiewicz: Mechanics of granular matter. Advances in applied mech., Vol. V, p. 233 (1958).

[83] C.N. DeSilva, K.A. Kline: Nonlinear constitutive equations for directed viscoelastic materials with memory. ZAMP 19, 128-139 (1968).

[84] O.W. Dillon, Jr.: A thermodynamic basis of plasticity. Acta Mechanica 3, 182-195 (1967).

[85] R.C. Dixon, A.C. Eringen: A dynamical theory of polar elastic dielectrics, 1,2. Int. J. Engng. Sci. 3, 359-298 (1965).

[86] S. Djurić: Dynamics of continua with internal orientation and its small vibrations.(In Serbocroatian) Doc. Diss., Belgrade Univ. (1964). Publ. Fac. Mech. Engng. - Section in Kragujevac, 1968.

[87] S. Djurić: A principle of virtual work for an oriented continuum and the problem of boundary conditions. Tehnika 7, 138 (1968).

[88] S. Djurić: On the plastic flow of Cosserat continua. (In Serbocroatian). To appear in Materials and Structures, Belgrade (1969).

[89] J.M. Doyle: On completeness of stress functions in elasticity. Tans. ASME (J. Appl. Mech.) 31, 728--729 (1964).

[90] J.M. Doyle: An extension of Green's method in elasticity, Proc. Konikl. Nederl. Akad. Wet. V 68 284-292 (1965).

[91] J.M. Doyle: Singular solutions in elasticity. Acta Mechanica 4, 27-33 (1967).

[92] T.C. Doyle, J.L. Ericksen:Nonlinear elasticity. Adv. Appl. Mech. 4, 53-115 (1956).

[93] J. Duffy, R.D. Mindlin: Stress-strain relations and vibrations of a granular medium. J. Appl. Mech., 24, (1957) pp. 585-593.

[94] P. Duhem: Le potentiel thermodynamique et la pression hydrostatique. Ann. Ecole Norm. 10, 187-230 (1893).

[95] G. Duvaut: Application du principle de l'indifférence matérielle à un milieu élastique matériellement polarisée. C.R. Acad. Sci. Paris 258, 3631-3634 (1964).

[96] U.B.C.O. Ejike: The plane circular crak problem in the linearized couple-stress theory. Int. J. Engng. Sci 7, 947-961 (1969).

[97] R.W. Ellis, C.W. Smith: A thin-plate analysis and experimental evaluation of couple-stress effects. Experimental Mechanics 7, 372 (1967).

[98] M.E. Erdogan, I.T. Gürgöze: Generalized Couette flow of a dipolar fluid between two parallel plates.ZAMP 20, 785-790 (1969).

[99] J.L. Ericksen: Note concerning the number of directions which, in a given motion, suffer no instantaneous rotations. J. Wash. Acad. Sci. 45, 65-66 (1955).

[100] J.L. Ericksen: Tensor Fields. Handbuch der Physik Bd. III/1 (ed. S. Fluegge), pp. 794-858. Berlin (1960).

[101] J.L. Ericksen: Transversely isotropic fluids. Kolloid--Zeitschrift 173, 117-122 (1960).

[102] J.L. Ericksen: Anisotropic fluids. Arch. Rat. Mech. Anal. 4, 231-237 (1960).

[103] J.L. Ericksen: Vorticity effects in anisotropic fluids. J. Polymer Sci. 47, 327-331 (1960).

[104] J.L. Ericksen: Theory of anisotropic fluids. Trans. Soc. Rheol. 4, 29-39 (1960).

[105] J.L. Ericksen: Conservation laws for liquid crystals. Trans. Soc. Rheol. 4, 29-39 (1960).

[106] J.L. Ericksen: Poiseuille flow of certain anisotropic fluids. Arch. Rat. Mech. Anal. 8, 1-8 (1961).

[107] J.L. Ericksen: Orientation induced by flow. Trans. Soc. Rheol. 6, 275-291 (1962).

[108] J.L. Ericksen: Kinematics of macromolecules. Arch. Rat. Mech. Anal. 9, 1-8 (1962).

[109] J.L. Ericksen: Hydrostatic theory of liquid crystals. Arch. Rat. Mech. Anal. 9, 371-378 (1962).

[110] J.L. Ericksen : Nilpotent energies in liquid crystal theory. Arch. Rat. Mech. Anal. 10, 189-196 (1962).

[111] J.L. Ericksen: Instability in Couette flow of anisotropic fluids. Quart. J. Mech. Appl. Math. 19, 455-459 (1966).

[112] J.L. Ericksen: Some magnetohydrodynamic effects in liquid crystals. Arch. Rat. Mech. Anal. 23, 266-275 (1966)

[113] J.L. Ericksen: Inequalities in liquid crystal theory. The Physics of Fluids 9, 1205-1207 (1966).

[114] J.L. Ericksen: Twisting of liquid crystals. J. Fluid Mech. 27, 59-64 (1967).

[115] J.L. Ericksen: General solutions in the hydrostatic theory of liquid crystals. Trans. Soc. Rheology 11, 5-14 (1967).

[116] J.L. Ericksen: Continuum theory of liquid crystals. Appl. Mech. Reviews 20, 1029-1032 (1967).

[117] J.L. Ericksen: Twisting of partially oriented liquid crystals. Quart. Appl. Math. 25, 474-479 (1968).

[118] J.L. Ericksen: Twisting of liquid crystals by magnetic fields. ZAMP 20, 383-388 (1969).

[119] J.L. Ericksen: Uniformity in shells. Arch. Rat. Mech. Anal. 37, 73-84 (1970).

[120] J.L. Ericksen: Simpler static problems in nonlinear theories of rods. Int. J. Solids Structures 6, 371-377 (1970).

[121] J.L. Ericksen, C. Truesdell: Exact theory of stress and strain in rods and shells. Arch. Rat. Mech. Anal. 1, 295-323 (1958).

[122] A.C. Eringen: Nonlinear theory of continuous media. McGraw-Hill Book Co. New York (1962).

[123] A.C. Eringen: Mechanics of micromorphic media. Applied Mechanics. Proc. XI Congress on Appl. Mech., München pp. 132-138 (1964).

[124] A.C. Eringen: Simple microfluids. Int. J. Engng. Sci 2, 205-217 (1964).

[125] A.C. Eringen: Theory of micropolar continua. Proc. 9th Midwestern Mech. Conference, Madison, Wisconsin (1965).

[126] A.C. Eringen: Linear theory of micropolar elasticity. J. Math. Mech. 15, 909-924 (1966).

[127] A.C. Eringen: Theory of micropolar fluids. J. Math. Mech. 16, 1-18 (1966).

[128] A.C. Eringen: Theory of micropolar plates. J. Appl. Math. Phys. 18, 12-30 (1967).

[129] A.C. Eringen: Linear theory of micropolar viscoelasticity. Int. Journal Engng. Sci. 5, 191-204 (1967).

[130] A.C. Eringen: Mechanics of micromorphic continua. Mechanics of Generalized Continua. Proc. IUTAM Symposium Freudenstadt-Stuttgart 18-35 (1967).

[131] A.C. Eringen: Mechanics of continua. John Wiley and Sons. Inc. New York (1967).

[132] A.C. Eringen: Theory of micropolar elasticity. Fracture, Vol. II, pp. 621-729. Academic Press, Inc. New York and London, (1968).

[133] A.C. Eringen: Micropolar fluids with stretch. Int. J. Engng. Sci. 7, 115-127 (1969).

[134] A.C. Eringen: Compatibility conditions of the theory of micromorphic elastic solids. J. Math. Mech. 19, 473-481. (1969).

[135] A.C. Eringen: On a theory of general relativistic thermodynamics and viscous fluids. Tech. Rep. N°8, Princeton Univ., Dept. of Aerospace and Mechanical Sciences pp. 1-34, March 1969.

[136] A.C. Eringen: Balance laws of micromorphic mechanics. Reprint (1970).

[137] A.C. Eringen, J.D. Ingram: A continuum theory of chemically reacting media. Int. J. Engng. Sci 3, 197-212 (1965).

[138] A.C. Eringen, E.S. Suhubi: Nonlinear theory of simple microelastic solids, Int. J. Engng. Sci. 2 189-203 (1964).

[139] A.C. Eringen, E.S. Suhubi: Stress distribution at two normally intersecting cylindrical shells. Nuclear structural Engng. 2, 253-270 (1965).

[140] R.L. Fodstick, K.W. Schuler: On Ericksen's problem for plane deformations with uniform transverse stretch. Int. J. Engng. Sci. 7, 217-233 (1969).

[141] N. Fox: A continuum theory of dislocations for polar elastic materials. Quart. Journ. Mech. Appl. Math. 19, 343-355 (1966).

[142] H. Frackiewicz: A plane problem of the theory of elasticity for media with a discrete lattice structure. Arch. Stosow. 19, 725-744 (1967).

[143] H. Frackiewicz: The bending problem of plane grates of discrete structure. Arch. Mech. Stosow. 22, 127--147 (1970).

[144] N.N. Frank: On the theory of liquid crystals. Discussions of Faraday Soc. 25, 19-28 (1958).

[145] D. Galletto: Sistemi incomprimibili a trasformazioni reversibili nel caso asimmetrico, I. Rend. Sem. Mat. Univ. Padova 36, 243-276 (1966).

[146] D. Galletto: On continuous media with contact couples. Meccanica 1, 18-27 (1966).

[147] M. Gligoric, R. Stojanovic: On the constitutive equations of statics of elastic dielectrica. (In Serbocroatian). VIII Yugoslav Congress of Mechanics, Split 1966. To appear in Tehnika.

[148] B.V. Gorskii: Nekatorie zadachi teorii nesimmetricheskoi uprugosti. Prkl. Mat. Mech. 3, 74-83 (1967).

[149] K.F. Graff, Y.H. Pao: The effects of couple-stresses on the propagation and reflection of plane waves in an elastic half-space. Journ. of Sound and Vibration 6, 217-229. (1967).

[150] A.E. Green: Anisotropic simple fluids. Proc. Roy. Soc. London (A), 279, 437-445 (1964).

[151] A.E. Green: Micro-materials and multipolar continuum mechanics. Int. J. Engng. Sci. 3, 533-537 (1965).

[152] A.E. Green, R.J. Knops, N. Laws: Large deformations, superposed small deformations and stability of elastic rods. Int. J. Solids Structures 4, 555-577 (1968).

[153] A.E. Green, N. Laws: A general theory of rods. Proc. Roy. Soc. A, 293, 145-155 (1966).

[154] A.E. Green, N. Laws: On the formulation of constitutive equations in thermodynamical theories of continua. Quart. J. Mech. Appl. Math. 20, 265-275 (1967).

[155] A.E. Green, N. Laws: A general theory of rods. Mechanics of Generalized Continua, Proc. IUTAM Symposium Freudenstadt-Stuttgart 1967, pp. 49-56.

[156] A.E. Green, N. Laws, P.M. Naghdi: A linear theory of straight elastic rods. Arch. Rat. Mech. Anal. 25, 285-298. (1967).

[157] A.E. Green, N. Laws, P.M. Naghdi: Rods, plates and shells. Proc. Camb. Phil. Soc. 64, 895-913 (1968).

[158] A.E. Green, B.C. McInnis, P.M. Naghdi: Elastic-plastic continua with simple force dipoles. Int. J. Engng. Sci. 6, 373-394 (1968).

[159] A.E. Green, P.M. Naghdi: A general theory of an elastic-plastic continuum. Arch. Rat. Mech. Anal. 18, 251 (1965).

[160] A.E. Green, P.M. Naghdi: Plasticity theory and multipolar continuum mechanics. Mathematika 12, 21-26 (1965).

[161] A.E. Green, P.M. Naghdi: On the derivation of discontinuity conditions in continuum mechanics. Int. J. Engng. Sci 2, 621-624 (1965).

[162] A.E. Green, P.M. Naghdi: A dynamical theory of interacting continua. Internat. J. Engng. Sci 3, 231-241 (1965).

[163] A.E. Green, P.M. Naghdi: Micropolar and director theories of plates. Quart. J. Mech. Appl. Math. 20, 183-199 (1967).

[164] A.E. Green, P.M. Naghdi: The linear theory of an elastic Cosserat plate. Proc. Camb. Phil. Soc. 63, 537--550 (1967).

[165] A.E. Green, P.M. Naghdi: The Cosserat surface. Mechanics of of Generalized Continua. Proc. IUTAM Symposium Freudenstadt-Stuttgart 36-48 (1967).

[166] A.E. Green, P.M. Naghdi, A note on the Cosserat surface. Quart. J. Mech. Appl. Math. 21, 135-139 (1968).

[167] A.E. Green, P.M. Naghdi: The linear elastic Cosserat surface and shell theory Int. J. Solids Structures 4, 585-592 (1968).

[168] A.E. Green, P.M. Naghdi: Shells in the light of generalized continua. Theory of Thin shells, Proc. 2nd IUTAM Symp., Copenhagen (1967) (ed. F.I. Niordson) pp. 39-58. Springer (1969).

[169] A.E. Green, P.M. Naghdi: Non-isothermal theory of rods, plates and shells. Int. J. Solids Structures 6, 209-244 (1970).

[170] A.E. Green, P.M. Naghdi, R.S. Rivlin: Directors and multipolar displacements in continuum mechanics. Int. J. Engng. Sci 2, 611-620 (1965).

[171] A.E. Green, P.M. Naghdi, W.L. Wainwright: A general theory of a Cosserat surface.Arch. Rat. Mech. Anal. 20, 287-308 (1965).

[172] A.E. Green, R.S. Rivlin: Simple forces and stress multipoles. Arch. Rat. Mech. Anal. 16, 325- 353 (1964).

[173] A.E. Green, R.S. Rivlin: Multipolar continuum mechanics. Arch. Rat. Mehc. Anal. 17, 113-147 (1964).

[174] A.E. Green, R.S. Rivlin: Multipolar continuum mechanics: functional theory part. 1, Proc. Roy. Soc. London (A), 284, 303-324 (1965).

[175] A.E. Green, R.S. Rivlin: The relation between director and multipolar theories in continuum mechanics. ZAMP 18, 208-218 (1967).

[176] A.E. Green R.S. Rivlin: Generalized continuum mechanics. Proc. IUTAM Symposium, Vienna 1966 pp. 132-145.

[177] A.E. Green, T.R.Steel: Constitutive equations for interacting continua. Int. J. Engng. Sci. 4, 483-500. (1966).

[178] J.M. Grigorenko: Ob antisimmetricheskom naprjazhenom sostojanju konicheskoi obolochki lineino peremenoi tolshchnini. Trudi konferencii po teorii plastin i obolochek, Kazan (1960) pp. 142-148.

[179] G. Grioli: Discontinuity wave and asymmetrical elasticity. Atti Accad. Naz. Lincei, R.C. Cl. Sci. Fis. Mat. Nat. 29, 309-312 (1960).

[180] G. Grioli: Elasticità asimmetrica. Ann. di Mat. Pura ed Applic. ser. IV, 50, 389-417 (1960).

[181] G. Grioli: Mathematical Theory of elastic equilibrium. (Recent Results) Ergebnisse der angew. Mathematik 7, Berlin-Göttingen-Heidelberg, (1962).

[182] G. Grioli: Elasticità asimmetrica. Applied Mechanics. pp. 252-254 Elsevier Publ. Co. Amsterdam-New York, 1962.

[183] G. Grioli: Sulla meccanica dei continui a trasformazioni reversibili con caratteristiche di tensione asimmetriche. Seminari dell'Istituto Nazionale di Alta matematica 1962/63, pp. 535-555 (1964).

[184] G. Grioli: On the thermodynamic potential for continuums with reversible transformations - some possible types. MECCANICA 1, 15-20 (1966).

[185] G. Grioli: On the thermodynamic potential of Cosserat continua. Mechanics of Generalized continua. Proc. IUTAM Symposium Freudenstadt-Stuttgart (1967), pp. 63-68.

[186] G. Grioli: Questioni di compatibilità per i continui di Cosserat. Symposia mathematica 1, 271-287 (1968).

[187] R.A. Grot: Thermodynamics of continuum with microstructure. Int. J. Engng. Sci 7, 801-814 (1969).

[188] R.A. Grot, J.D. Achenbach: Large deformations of a laminated composite. Int. J. Solids Structures 6, 641-659 (1970).

[189] W. Günther: Zur Statik und Kinematik des Cosseratschen Kontinuums. Abh. Braunschw. Wiss. Ges. 10, 195-213 (1958).

[190] W. Günther: Analoge Systeme von Schalengleichungen. Ingenieur-Arch. 30, 160-186 (1961).

[191] B.P. Guru: Concept of couple-stress theory. J. Scient. Res. Banaras Hindu Univ. 15, 36-50 (1964-65).

[192] A.G. Guz, C.N. Savin: The plane problem of couple-stress theory of elasticity for an infinite plane weakened by a finite number of circular holes. (In Russian) Prikl. Math. Mekh. 30, 1043 (1967).

[193] R.J. Hartranft, G.C. Sih: The effect of couple-stresses on the stress-concentration of a circular inclusion. Trans. ASME E32 (J. Appl. Mech.), 429-431 (1965).

[194] R.J. Hartranft, G.C. Sih: Discussion on the paper " Couple-stress effects on stress concentration around a cylindrical inclusion in a field of uniaxial tension" by Y. Weitsman, Trans. ASME E 32 954-956 (1965).

[195] R.J. Hartranft, G.C. Sih: Uniqueness of the concentrated-load problem in the linear theory of couple-stress elasticity. Meccanica 3, 195-198 (1968).

[196] R.J. Hartranft, G.C. Sih: Effect of plate thickness on the bending stress distribution around through cracks. J. Math. Phys. 47, 276-291 (1968).

[197] R. Hayart: Extension des formaulas de Murnaghan relatives au solide en phase d'elasticité finie, au cas de couples superficiels. C.r. Acad. Sci. Paris, 258, 1390-1391 (1964).

[198] G. Hazen, Y. Weitsman: Stress concentration in strain--gradient bodies. J. Engng. Mech. Div. Proc. ASCE 94, 773-795 (1968).

[199] F. Hehl, E. Kröner: Über den Spin in der allgemeinen Relativitätstheorie: Eine notwendige Erweiterung der Einsteinschen Feldgleichungen. Z. Phys. 187, 478 (1965).

[200] F. Hehl, E. Kröner: Zum Materialgesetz eines elastischen Mediums mit Momentenspannungen. Z. Naturf. 20 a 336-350 (1965).

[201] K. Hellan Jr.: Rotation moments in continuous structures Tekniske Skr. 18 N 42-44 (1958).

[202] E. Hellinger: Die allgemeinen Ansätze der Mechanik der Kontinua. Enz. Math. Wiss. 4, 602-694 (1914).

[203] H. Herrmann, J.D. Achenbach: Applications of theories of generalized Cosserat continua to dynamics of composite materials. Mechanics of Generalized Continua. Proc. of the IUTAM Symposium Freudenstadt-Stuttgart 1967 pp. 69-79.

[204] R. Hill: The essential structure of constitutive laws for metal composites and polycrystals. J. Mech. Phys. Solids 15, 79-95 (1967).

[205] R.N. Hills: The slow flow of a dipolar fluid past a sphere. Int. J. Engng. Sci. 5, 957 (1967).

[206] I. Hlavacek, M. Hlavacek: On the existence and uniqueness of solution and some variational principles in linear theories of elasticity with couple-stresses. Aplikace Matematiki 14, 387-410; 411-427 (1969).

[207] O. Hoffmann: On bending of thin elastic plates in the presence of couple-stresses. Trans. ASME E 31 (J. Appl. Mech.), 706-707 (1964).

[208] W.P. Hoppmann, F.O.F. Shahwan: Physical model of a three-constant isotropic elastic material Tans. ASME 32 E (J. Appl. Mech.) 4, 837-841, (1965).

[209] C.C. Hsiao, T.S. Wu: Orientation and Strength of Branched Polymer systems. J. Polymer Sci. 1, 1789-1797 (1963).

[210] C.L. Huang: Wave propagation in an isotropic medium with couple stresses. Trans. ASME E 33 (J. Appl. Mech) 939-941 (1966).

[211] C.L. Huang: The energy function for anisotropic materials with couple-stresses-cubic and hexagonal systems. Int. J. Engng. Sci. 6, 609-621 (1968).

[212] C.L. Huang: The energy function for crystal materials with couple-stresses. Int. Engng. Sci. 7, 1221-1229 (1969).

[213] C.L. Huang, G.F. Smith: The energy function for isotropic materials with couple stresses. ZAMP 18, 905-909 (1967).

[214] B. Hudimoto, T. Tokuoka: Two-dimensional shear flows of linear micropolar fluids. Int. J. Eneng. Sci. 7, 515-522, (1969).

[215] R.R. Huilgol: On the concentrated force problem for two-dimensional elasticity with couple-stresses. Int. J. Engng. Sci. 5, 81-93 (1967).

[216] D. Iesan: On the linear theory of micropolar elasticity. Int. J. Engng. Sci. 7, 1213-1220 (1969).

[217] E.A. Il'iushina: On a model of continuous medium, taking into account the microstructure. Appl. Math. Mech. 33, 896-902 (1969)(Translated from PMM 33, 917-923) (1969).

[218] J.D. Ingram, A.C. Eringen: A continuum theory of chemically reacting media -I. A.C. Eringen, J.D. Ingram: A continuum theory of chemically reacting media -II. Int. J. Engng. Sci. 3, 197 (1965); 5, 289-322 (1967).

[219] D.E. Johnson: A difference-based variational method for shells. Int. J. Solids Structures 6, 699-724 (1970).

[220] A.I. Kalandja, V.S. Zhgenti: O ploskih zadach momentnoi teorii uprugosti. Tbilissk Matem. In-ta AN Gruz. SSR 33, 57-75 (1967).

[221] S. Kaliski: On a model of the continuum with an essentially non-symmetric tensor of mechanic stress. Arch. Mech. Stosow. 15, 33-45 (1963).

[222] S. Kaliski, Z. Plohocki, D. Rogula: The asymmetry of the stress tensor and the angular momentum conservation for a combined mechanical and electromagnetic field in a continuous medium. Bull. Acad. Pol. Sci. Sér. Sci. Techn. 10, 189–195 (1962).

[223] P.N. Kaloni: On certain steady flows of anisotropic fluids. Int. J. Engng. Sci. 3, 515–532 (1965).

[224] P.N. Kaloni: Certain periodic flows of anisotropic fluids. Physics of Fluids. 9, 1316–1321 (1966).

[225] P.N. Kaloni, T. Ariman: Stress concentration effect in micropolar elasticity. ZAMP 18, 136–141 (1967).

[226] U. Kammerer: Elastische Deformationen in Supraleitern. (Dissertation, Teil 1). Z. Physik 227, 125–140. (1969).

[227] E.F. Keller, J.B. Keller: Statistical mechanics of the moment stress tensor. Physics of Fluids, 9, 1, 3–7 (1966).

[228] E.F. Keller: Statistical mechanics of a fluid in an external potential. The Phys. Rev. 142, 90–99 (1966).

[229] S. Kessel: Linear theory of elasticity of anisotropic Cosserat continuum. Abh. Braunschw. Wiss. Ges. 16 1–22 (1964).

[230] S. Kessel: Conditions of compatibility in a Cosserat continuum. Abh. Braunschw. Wiss. Ges. 17, 51–61 (1965).

[231] S. Kessel: Stress functions and loading singularities for the infinitely extended, linear elastic-isotropic Cosserat continuum. Mechanics of Generalized Continua. Proc. IUTAM Symposium Freudenstadt -Stuttgart (1967), pp. 114–119.

[232] A.M. Kirichenko: Uchet momentnih naprjazhenij pri isledovanii osesimmetrichnogo naprjazhenogo sostojania uprugih sferi. Prikl. Mekhanika 4, 47-54 (1986).

[233] A.D. Kirwan, Jr., N. Newman: Plane flow of a fluid containing rigid structures. Int. J. Engng. Sci. 7, 883-894 (1969).

[234] A.D. Kirwan, N. Newman: Simple flow of a fluid containing deformable structures. Int. J. Engng. Sci. 7, 1067-1078 (1969).

[235] K.A. Kline: On thermodynamics of directed continuous media. ZAMP 19, 793 (1968).

[236] K.A. Kline, S.J. Allen: Heat Conduction in Fluids with Substructure. ZAMM 48, 435-444,, (1968).

[237] K.A. Kline, S.J. Allen: On continuum theories of suspensions of deformable particles. ZAMP 19, 898-905 (1969).

[238] K.A. Kline, S.J. Allen: Fluid suspensions: Flow formulation in Couette motion. ZAMP 21, 26-36 (1970).

[239] K.A. Kline, S.J. Allen, C.N. DeSilva: A continuum approach to blood flow. Biorheology. 5, 111-118. (1968).

[240] G. Kluge: Zur Dynamil der allgemeinen Versetzungstheorie bei Berücksichtigung von Momentspannungen. Int. J. Engng. Sci. 7, 169-182 (1969).

[241] W.T. Koiter: Couple-stresses in the theory of elasticity, I, II Proc. Konikl. Nederl. Akad. Wet. 67, 17-29, 30-44 (1969).

[242] 1) W.T. Koiter: A comparison between John's refined interior shell equations and classical shell theory. ZAMP 20, 642-652 (1969).

2) F. John: Appendix to the preceeding paper by W.T. Koiter. ibid. 652-653.

[243] S. Komljenović: Plastic flow with the non-symmetric stress tensor. (In Serbocroatian). Doct. Diss. Belgrade Univ. 1964. Publ. Fac. Mech. Engng. - Section Kragujevac 10,109-145 (1966).

[244] I. Kozak: On equilibrium equations of solid continua. Acta Technica 59, 141-163. (1967).

[245] H. Kronmüller, H. Riedle: Description of elastic and dielastic interactions in superconductors by quasidislocations. Phys. Stat. Sol. 38, 403-407 (1970).

[246] E. Kröner: Kontinuumstheorie der Versetzungen und Eigenspannungen. Ergeb. Angew. Math. 5, 1-179. Berlin--Göttingen-Heidelberg, 1958 .

[247] E. Kröner, A. Seeger: Nicht-lineare Elastizitätstheorie der Versetsungen und Eigenspannungen. Arch. Rat. Mech. Anal. Anal. 3, 97-119 (1959).

[248] E. Kröner: Allgemeine Kontinuumstheorie der Verstetzungen und Eigenspannungen. Arch. Rat. Mech. Anal. 4, 273-334 (1960).

[249] E. Kröner: Die neuen Konzeptionen der Kontinuumsmechanik der festen Körper. Phys. Stat. Sol. 1, 3-16 . (1961).

[250] E. Kröner: The dislocation as a fundamental new concept in continuum mechanics. Mater. Sci. Research 1, 281 (1962).

[251] E. Kröner: Dislocations and continuum mechanics. Appl. Mech. Reviews 15, 599-606 (1962).

[252] E. Kröner: On the physical reality of torque-stresses in continuum mechanics. Int. J. Engng. Sci. 1, 261-278.(1963).

[253] E. Kröner: Das physikalische Problem der antisymmetrischen Spannungen und der sogenannten Momentspannungen. Applied Mechanics, Proc. 11th IUTAM Congress, München 1964, pp. 143-150.

[254] E. Kröner: Elasticity theory of materials with long range cohesive forces. Int. J. Solids Structures 3, 731-742 (1967).

[255] E. Kröner: The mechanics of generalized continua: Physical foundations based upon the rigorous discrete particle mechanics. Symposia Mathematica 1, 313-326 (1969).

[256] E. Kröner: Über die Symmetrieeigenschaften elastischer Materiälen mit Kohäsionskräften endlicher Reichweite. Problemi Gidrodinamiki i Mekaniki Splosnoi sredi. Sedov Aniversary Volume, Moskva 1969. Izdateljstvo "Nauka" pp. 293-300.

[257] I.A. Kunin: Field of arbitrarily distributed dislocations and force dipoles in an unbounded elastic anisotropic medium. Zb. Prikl. Mekh. Tekh. Fiz. (PMTF) 5, 76-83 (1965).

[258] I.A. Kunin: The theory of elastic media with microstructure and the theory of dislocations. Mechanics of Generalized continua. Proc. IUTAM Symposium Freudenstadt-Stuttgart 1967, pp. 321-329.

[259] I.A. Kunin: Vnutrenie naprjazenija v srede s mikrostrukturoj. Prikl. Mat. Meh. 31, 889-896 (1967).

[260] E.V. Kuvshinskii, E.L. Aero: Kontinual'naja teorija asimmetricheskoi uprugosti. Uchet "vnutrennego" vrashchenija. Fiz. tv. tela 5, 2591-2598 (1963).

[261] E.L. Kyser: Linear elastic dipolar plates. J. Appl. Mech. Trans. ASME E 35, 499-504 (1968).

[262] S.K. Lakshmana Rao, N.C. Pattabhi Ramacharyulu, P. Bhujanga Rao: Slow steady rotation of a sphere in a micropolar fluid. Int. J. Eneng. Sci. 7, 905-916. (1969).

[263] S.H. Lam, D.C. Leigh: Inner-outer expansion method for couple-stresses. AIAA Journal, 2, 1511-1512 (1964).

[264] N. Laws: A simple dipolar curve. Int. J. Engng. Sci 5, 653 (1967).

[265] D.C. Leigh: On the restriction of processes by the Clausius-Duhem inequality. ZAMP 20, 167-176 (1969).

[266] M.Ya. Leonov, K.N. Rusinko: Macro-stress in elastic bodies. Zb. Prikl. Mekh. Tekh. Fiz. (PMTF) 1 104-110 (1963).

[267] F.M. Leslie: Some constitutive equations for anisotropic fluids. Quart. J. Mech. Appl. Math. 19, 357-370 (1966).

[268] F.M. Leslie: Some constitutive equations for liquid crystals. Arch. Rat. Mech. Anal. 28, 265-283 (1968).

[269] M. Levinson: The complementary energy theorem in finite elasticity. Trans. ASME 32 E (J. Appl. Mech.) 826-828 (1965).

[270] H. Lippmann: Eine Cosserat Theorie des plastischen Fliessens. To appear in Acta Mechanica.

[271] A.T. Listrov: Model of a viscous fluid with an antisymmetric stress tensor. PMM 31, 115 (1967).

[272] C.Y. Liu: On turbulent flow of micropolar fluids. Int. Sci. 8, 457-466 (1970).

[273] F.J. Lockett: On squires theorem for viscoelastic fluids. Int. J. Engng. Sci. 7, 337-349 (1969).

[274] V.V. Lohin: Obscie formi svjazi mezdu tenzornimi poljami v anizotropnoj srede, svojstva kataroj opisivajut sja vektorami, tenzorami vtarova ranga i antisimetricnimi tenzorami tretevo ranga. Dokl. ANSSSR 149, 1282-1285 (1963).

[275] V.A. Lomakin: Voprosi obobshchenoi. momentnoi teorii uprugosti. Vestn. Mosk. Univ.-Mat. Meh. 1, 82-88 (1967).

[276] V.A. Lomakin: On the theory of deformation of micrononhomogeneous bodies and its relation with the couple-stress theory of elasticity. PMM 30, 1035 (1967).

[277] E.F. Low, W.F. Chang: Stress concentrations around shaped holes. J. Engng. Mech. Division, Proc. ASCE 93, 33 (1967).

[278] M.F. McCarthy, A.C. Eringen: Micropolar viscoelasticity waves. Int. J. Engng. Sci. 7 447-458 (1969).

[279] J.A. McLennan: Symmetry of the stress tensor. Physika 32, 689 (1966).

[280] B. Meissonnier: On equilibrium figures of a hyperelastic line in an oriented medium. Journal de mecanique 4, 423-437 (1965).

[281] S. Minagawa: On the force on continuously distributed dislocations in a Cosserat continuum. RAAG Res. Notes (3rd. Series) 144, Oct. 1969.

[282] S. Minagawa: On the possibility of assuming stress-function spaces with reference to the completeness theorem for the Beltrami and Günther representations of stresses and moment-stresses. RAAG Res. Notes (3d sen) 151, pp. 1-14 (1970).

[283] R.D. Mindlin: Influence of couple-stresses on stress concentration. Exptl. Mech. 3, 1-7 (1963).

[284] R.D. Mindlin: Micro-structure in linear elasticity. Arch. Rat. Mech. Anal. 16, 51-78 (1964).

[285] R.D. Mindlin: On the equations of elastic materials with micro-structures. Int. J. Solids Structures 1, 73-78 (1965).

[286] R.D. Mindlin: Stress functions for a Cosserat continuum. Int. J. Solids Structures 1, 265-271 (1965).

[287] R.D. Mindlin: Second gradient of strain and surface tension in linear elasticity. Int. J. Solids and Structures, 1, 417-438 (1965).

[288] R.D. Mindlin: Theories of elastic continua and crystal lattice theories. Mechanics of Generalized continua. Proc. IUTAM Symposium Freudenstadt-Stuttgart 1967, pp. 312-320.

[289] R.D. Mindlin: Polarization gradient in elastic dielectrics. Int. J. Solids Structures 4 637-642 (1968).

[290] R.D. Mindlin: Continuum and lattice theories of influence of electromechanical coupling on capacitance of thin dielectric films. Int. J. Solids Structures 5, 1197-1208 (1969).

[291] R.D. Mindlin, N.N. Eshel: On first strain-gradient theories in linear elasticity. Int. J. Solids Structures 4, 109-124 (1968).

[292] R.D. Mindlin, H.F. Tiersten: Effects of couple-stresses in linear elasticity. Arch. Rat. Mech. Anal. 11, 415-448 (1962).

[293] M. Misicu: Theory of viscoelasticity with couple-stresses and some reductions to two-dimensional problems, I,II Rev. Roum. Sci. Techn. Mec. Appl. 8, 921-925 (1963); 9, 3-35 (1964).

[294] M. Misicu: On a theory of asymmetric plastic and visco-elastic-plastic solids. Rev. Roumaine Sci. Tech. Méc. Appl. 9, 477-495 (1964).

[295] M. Misicu: A Generalization of the Cosserat equations of the motion of deformable bodies (with internal degrees of freedom) Rev. Roum. Sci. Tech. Méc. Appl. 9, 1351-1359 (1964).

[296] M. Misicu: Asupra unei teorii, a plasticitatii si visco elasticitatii asimetrice. Studii si cercetari mec. apl. Acad. RPR, 17, 1073- 1091 (1964).

[297] M. Misicu: On a general solution of the theory of singular dislocations of media with couple-stresses. Rev. Roum. Sci. Techn., Méc. Appl. 10, 35- 46 (1965).

[298] M. Misicu: On the mechanics of structural media, non-correlated fields. Rev. Roum. Sci. Tech., Méc. Appl. 10, 295-331 (1965).

[299] M. Misicu: The elasticity of structural non-homogeneous centro-asymmetric isotropic bodies. Rev. Roum. Sci. Techn. Méc. Appl. 10, 1085-1104 (1965).

[300] M. Misicu: On the distorsions in special structural media. Rev. Roum. Sci. Techn. Méc. Appl. 10, 1441--1460.

[301] M. Misicu: A generalization of the Cosserat equations of the motion of deformable bodies. Arch. Mech. Stosow. 17, 183-190 (1965).

[302] M. Misicu: O extindere a ecuatiilor lui Cosserat relative la miscarea mediilor deformable structural neomegene. Studii si cercetari mec. appl. Acad. RSR 19, 839-847 (1965).

[303] M.Misicu: The structural distorsion and the basic laws of the structural dislocation theory. Rev. Roum. Sci. Techn., Méc. Appl. 11, 109–123 (1966).

[304] M. Misicu: The coupled mechanical model. Rev. Roumaine Sci. Techn. Méc. Appl. 12, 177–199 (1967).

[305] M. Misicu: The mechanics of heterogeneous media with independent coupling motions. Rev. Roumaine Sci. Techn. Méc. Appl. 12, 511–525 (1967).

[306] M. Misicu: Mécanica mediilor deformabile. Fundamentale elasticitatii structurale. Bucuresti (1967).

[307] M. Misicu: The mechanics of heterogeneous media with independent coupling motions. Rev. Roum. Sci. Tech. Méc. Appl. 13, 49–50 (1968).

[308] M. Misicu: Interpolation principles of dynamics and dual continuum theory of material systems. Rev. Roum. Sci. Techn. Méc. Appl. 13, 511–521 (1968).

[309] M. Misicu: The generalized dual continuum in elasticity and dislocation theory. Mech. Generalized Continua, Proc. IUTAM Symposium Freudenstadt-Stuttgart, 1967, pp. 141–151 (1968).

[310] M. Misicu: A theory of thin shells with local structure effects. Theory of thin Shells, Proc. 2nd. IUTAM Symp., Copenhagen 1967 (ed. F.I. Niordson) pp. 106–114 Springer 1969.

[311] M. Misicu; V. Nicolae: A distorsional type theory of thin shells. Rev. Roum. Sci. Techn. Méc. Appl. 13, 339–345 (1968).

[312] B. Morandi: Internal couples in solids and structural continua. (In Polish) Rozprawy Inzynierskie 14, 571–589 (1966).

[313] R. Muki, E. Sternberg: The influence of couple-stresses on singular stress concentrations in elastic solids. ZAMP 16, 611-648 (1965).

[314] M.G.K. Murthy: Thermal bending of polar orthotropic plates. Res. Bull. Regional Engng. College, Warangal 1, 82-87 (1964).

[315] N. Naerlovic-Veljkovic, R. Stojanovic: Finite thermal strains in a hollow circular cylinder (in Serbo-croatian) Tehnika 21, 134-140 (1966).

[316] N. Naerlovic-Veljkovic: Rotationally symmetric thermoe lastic stress in materials with couple-stresses. (In Serbocroatian) Proc. IX Yugoslav Congress of Mechanics Split 1968, pp. 137-146.

[317] N. Naerlovic-Veljkovic, R. Stojanovic, L. Vujosevic: Applications of the general theory of incompati ble deformations on thermoelasticity. (In Serbo croatian) Tehnika 24, 9-13 (1969).

[318] P.M. Naghdi: On the differential equations of the linear theory of elastic shells. Applied Mechanics. - Proc. 11th. IUTAM Congress, München 1964, pp. 262-269.

[319] P.M. Naghdi: A static-geometric analogy in the theory of couple-stresses. Proc. Ned. Acad. Wet. (B) 68, 29-32 (1965).

[320] P.M. Naghdi: A theory of deformable surface and shell theory. L.H. Donnell Anniversary volume, Proc. Symp. on the Theory of Shells, Houston, 1966, pp; 25-43.

[321] Ju. Nemish: Koncentracia naprjazhenia okolo krivolineinii otverstvii v nesimmetrichnoi teorii uprugosti. Prikl. Mekh. 2, 85-96 (1966).

[322] Ju. Nemish: Ploskaja zadacha momentnoi teorii uprugosti dla oblasti s krugovim otverstviem pri zadanih na konture peremeschchenijah. Dopovidi AN USSR 6, 748-753 (1966).

[323] Ju. Nemish, V.P. Tretyak: Flexure of a cantilever, weakened by a circular hole, considering couple stresses. Prikl. Mekh. 3, 131-134 (1967).

[324] H. Neuber: On the general solution of linear elastic problems in isotropic and anisotropic Cosserat continua. Applied. Mechanics. Proc. XI Internat. Congress of Appl. Mech. München 1964 pp. 153-158.

[325] H. Neuber: On problems of stress concentration in a Cosserat body. Acta Mechanica, 2, 48-69 (1966).

[326] H. Neuber: Statistische Stabilität nichtlinear elastischer Kontinua mit Anwendung auf Schalen. ZAMM 46, 211-220. (1966).

[327] H. Neuber: On the effect of stress concentration in Cosserat continua. Mechanics of Generalized Continua. Proc. IUTAM Symposium Freudenstadt-Stuttgart 1967.

[328] H. Neuber, H.G. Hahn: Stress concentration in scientific research and engineering. Appl. Mech. Rev. pp. 187, (1966).

[329] V.N. Nikolaevskii, E.F. Afanasiev: On some examples of media with microstructure of continuous particle. Int. J. Solids Structures 5, 671-678 (1969).

[330] W. Noll: A mathematical theory of mechanical behaviour of continuous media. Arch. Rat. Mech. Anal. 2 197-226 (1958).

[331] W. Noll: Materially uniform simple bodies with inhomogeneities. Arch. Rat. Mech. Anal. 27, 1-32 (1967).

[332] V.V. Novozilov: O svjazi mezdu naprjazenijami i uprugimi deformacijami v polikristallah. Problemi Gidromehaniki i Mehaniki Splosnoi Sredi. Sedov Anniversary Volume, Moskva 1969.Izd. "Nauka" pp. 365-376.

[333] W. Nowacki: Couple-stresses in the theory of thermoelasticity. Proc. IUTAM Symposium Vienna, 1966. pp. 259-278.

[334] W. Nowacki: Couple-stresses in the theory of thermoelasticity, I, II, III. Bull. Acad. Pol. Sci., Sci. Techn., 14, 97-106, 203-212, 505-513 (1966).

[335] W. Nowacki: Some theorems of asymmetric thermoelasticity. Bul. Acad. Pol. Sci., Sci. Techn., 289-296 (1967).

[336] W. Nowacki: On the completeness of stress functions in thermoelasticity. Bul. Acad. Pol. Sci., Sci. Techn. 15, 583-591 (1967).

[337] W. Nowacki: On the completeness of stress functions in asymmetric elasticity. Bull. Acad. Pol. Sci., Sci. techn. 16, 309-315 (1968).

[338] W. Nowacki: Naprezenia momentowe w termosprezystosci. Rozprawy Inzynierskie 16, 441-471 (1968).

[339] W. Nowacki: On the completeness of potentials in a micropolar elasticity. Arch. Mech. Stosowanej 21, 107-122 (1969).

[340] W. Nowacki: Formulae for overall thermoelasticity deformations in a micropolar body. Rev. Roumaine Sci. Techn. Méc. Appliquée 15, 269-276 (1970).

[341] W. Nowacki, W.K. Nowacki: Propagation of monochromatic waves in an infinite micropolar elastic plate. Bull. Acad. Polon. Sci., Sci. Techn. 17, 29-37 (1969).

[342] W. Nowacki, W.K. Nowacki: Propagation of elastic waves in a micropolar cylinder. I, II. Bull. Acad. Pol. Sci., Sci Techn. 17, 39-47, 49-56 (1969).

[343] W. Nowacki, W.K. Nowacki: Generation of waves in an infinite micropolar elastic solid body I, II. Bull. Acad. Polon. Sci., Sci. Techn. 17, 76-82, 83-90 (1969).

[344] W. Nowacki, W.K. Nowacki: The plane Lamb problem in a semi-infinite micropolar elastic body. Arch. Mech. Stosow. , 21, 241-261 (1969).

[345] J.L. Nowinski, I.A. Ismail: Certain two-dimensional problems concerning the Cosserat medium. Development in Mechanics. Proc. IX Midwestern Conference, Madison, Wisc. 1965, pp. 345-355.

[346] W.C. Orthwein: On finite strain-displacement relations. Tensor 17, 139-149 (1967).

[347] N. Oshima: Dynamics of granular media. RAAG Memoirs, Vol. 1, pp. 563-572. Tokyo 1955.

[348] N.J. Pagano: Stress singularities in the bending of Cosserat plates. Doct. Diss. Lehing Univ. (1967).

[349] N.J. Pagano, G.C. Sih: Load-induced stress singularities in the bending of Cosserat plates. Meccanica, 3, 34-42 (1968).

[350] N.J. Pagano, G.C. Sih: Stress singularities around a crack in a Cosserat plate. Int. J. Solids Structures 4, 531-553 (1968).

[351] V.A. Pal'mov: Fundamental equations of the theory of asymmetric elasticity. J. Appl. Math. Mech. 28, 496-505 (1964).

[352] V.A. Pal'mov: Ploskaja zadacha teorii nesimmetricheskoi uprugosti. PMM 28 117-1120 (1964).

[353] V.A. Pal'mov: On a model of a medium with complex structure. Appl. Math. Mech. 33, 747-753 (1969)(translated from PMM 33, 768-773) (1969).

[354] V.R. Parfit, A.C. Eringen: Reflection of plane waves from the flat boundary of a micropolar elastic half-space. J; Acoust. Soc. Amer. 45, 1258-1227 (1969).

[355] G. Paria: Effect of Cosserat's couple-stresses in the stress distribution in a semi-infinite medium with varying modulus of elasticity. J. Australian Math. Soc. 6, 157-171 (1966).

[356] J.J.C. Picot, A.G. Fredrickson: Interfacial and electrical effects on thermal conductivity of nematic liquid crystal. I & EC, Fundamentals 7, 84-89 (1968).

[357] F. Pietras, J. Wirwinski: Termal stresses in a plane anisotropic Cosserat continuum. Arch. Mech. Stosow. 19, 627-635 (1967).

[358] M. Plavsić: Thermodynamics of viscous flow with the non-symmetric stress tensor. (In Serbocroatian) Magisterial Diss. Belgrade Univ. (1965).

[359] M. Plavsić: Viscometric flow of fluids with the non-symmetric stress tensor. (In Serbocroatian) Doc. Diss. Belgrade Univ. 1966.

[360] M. Plavsić: On the influence of couple stresses on the distribution of velocities in the flow of polar fluids. Mechanics of Generalized Continua. Proc. IUTAM Symposium Freudenstadt-Stuttgart 1967, pp. 160-161.

[361] M. Plavsić: Constitutive equations for dipolar fluids. To appear.

[362] M. Plavsić, D. Radenković.:Viscoplastic flow with the non-symmetric stress tensor. To appear.

[363] M. Plavsić, R. Stojanović: On constitutive equations for viscous fluids with couple-stresses. (In Serbo-croatian) Naucno-tehnicki pregled 16, N°7, 3-12 (1966).

[364] Yu.N. Podilchuk, Kirichenko: The effect of spherical cavity upon the state of stress in an elastic medium, with consideration of couple-stress. Prikl. Mekh. 3, 69-79 (1967).

[365] J.W. Provan, R.C. Köller: On the theory of elastic plates. Int. J. Solids Structures 6, 933-950 (1970).

[366] S. Ranković: Konstante naponskih spregova. Referat I/22 na Kongresu SJL, Ljubljana,Novembra 1969.

[367] S.K.L. Rao, N.C.P. Ramacharyulu, P.B. Rao: Slow steady rotation of a sphere in a micropolar fluid. Int. J. Engng. Sci. 7, 905-916 (1969).

[368] E. Reissner: On the foundations of the theory of elastic shells. Applied Mechanics,Proc. XI. Internat. Congress Appl. Mech. München 1964, pp. 20-30.

[369] E. Reissner: On St. Venant's flexure including couple-stresses. PMM, Appl. Math. Mech. 32, 944-950 (1968).

[370] E. Reissner: On the foundation of generalized linear shell theory. Theory of thin shells. Proc. 2nd IUTAM Symp. Copenhagen, 1967 pp. 15-30, Springer 1969 (ed. F.I.Niordson).

[371] E. Reissner: On the equations of nonlinear shallow shell theory. Studies in Appl. Math. 48, 171-175 (1969).

[372] E. Reissner: On generalized two-dimensional plate theory I, II., Int. J. Solids Structures 5, 525-532, 629-637 (1969).

[373] E. Reissner: On axially uniform stress and strain in axially homogeneous cylindrical shells. Int. J. Solids Structures 6, 133-138 (1970).

[374] E. Reissner, F.Y.M. Wan: A note on Günther's analysis of couple stress. Mechanics of Generalized Continua. Proc. IUTAM Symposium Freudenstadt-Stuttgart 1967 pp. 83-86.

[375] E. Reissner, F.Y.M. Wan: Rotationally symmetric stress and strain in shells of revolution. Studies in Appl. Math. 48, 1-17 (1969).

[376] E. Reissner, F.Y.M. Wan: On the equations of linear shallow shell theory. Studies in Appl. Math. 48, 133-145 (1969).

[377] R.S. Rivlin: Generalized mechanics of continuous media. Mechanics of Generalized Continua Proc. IUTAM Symposium Freudenstadt-Stuttgart 1967 pp. 1-17.

[378] R.S. Rivlin: The formulation of theories in generalized continuum mechanics and their physical significance. Symposia Mathematica 1, 357-373 (1969).

[379] C. Rymarz: On the model of nonsimple medium with rotation degrees of freedom. Bull. Acad. Polon. Sci. Ser. Sci. Techn. 16, 271-277 (1968).

[380] M.A Sodowsky, Y.C. Hsu, M.A. Hussain: Boundary layers in couple-stress elasticity and stiffening of thin layers in shear, Matervliet Arsenal, Benet R. a. E. Labs. (Matervliet, N.Y.) WVT-RR 6320, 56 pp. Oct. (1963).

[381] M.A. Sodowsky, Y.C. Hsu, M.A. Hussain: Effect of couple-stresses on force transfer between embedded microfibers. Journal of composite materials, 1, 174-187 (1967).

[382] N. Sandru: Reciprocity theorem in asymmetric elasticity (dynamic case) Atti Accad. Naz. Lincei, Rendiconti Cl. Sci. Fis. Mat. Nat. 38, 78-81 (1965).

[383] N. Sandru: The reciprocity theorem of the Betti type in asymmetric elasticity. Compt. Rend. Acad. Sci. Paris 260, 3565-3567 (1965).

[384] N. Sandru: On some problems of the linear theory of the asymmetric elasticity. Int. J. Engng. Sci. 4, 81-94 (1966).

[385] M. Satake: Some considerations on the mechanics of granular media. Mechanics of Generalized Continua. Proc. IUTAM Symposium Freudenstadt-Stuttgart 1967, pp. 156-159.

[386] M. Satake: Mechanical quantities and their relations in continuum and discrete materials. Preprint, Roum. Nat. Conference on Appl. Mechanics, Bucarest, June, 1969.

[387] G.N. Savin: Raspredelenie naprjazhenija okolo otverstii. Kiev. 1968.

[388] G.N. Savin, A.N. Guz: A method of solution of plane problems in couple-stress elasticity for multiconnected regions. Prikl. Mekh. 11, 3-19 (1966).

[389] A. Sawczuk: On yielding of Cosserat Continua. Arch. Mech. Stosow. 19, 471-480 (1967).

[390] H. Schäfer: Versuch einer Elastizitätstheorie des zweidimensionalen ebenen Cosserat-Kontinuums. Miszellaneen der Angewandten Mechanik. Festschrift W. Tolmien, 277-292, Berlin 1962.

[391] H. Schäfer: Continui di Cosserat. Funzioni potenziali. Calcolo numerico delle piastre. Cemento 63, 287-296, 327-336, 369-375 (1966).

[392] H. Schäfer: Analysis der Motorfelder im Cosserat-Kontinuum. ZAMM 47, 319 (1967).

[393] H. Schäfer: Das Cosserat-Kontinuum. ZAMM 47, 485-498 (1967).

[394] H. Schäfer: Die Spannungsfunktionen eines Kontinuums mit Momentspannungen. I,II. Bull. Acad. Pol. Sci. Ser. Sci. Techn. 15, 71-75, 77-81 (1967).

[395] H. Schäfer: The basic affine connections in a Cosserat continuum. Mechanics of Generalized Continua. Proc. IUTAM Symposium Freudenstadt-Stuttgart 1967. pp. 57-62.

[396] H. Schäfer: Eine Feldtheorie der Versetzungen in Cosserat-Kontinuum. ZAMP 20, 891-899. (1969).

[397] H. Schäfer: Maxwell-Gleichungen bewegter Versetzungen im Cosserat Kontinuum. To appear.

[398] H. Schäfer: Maxwell-Gleichungen, Energiesatz und Lagrange-Dichte in der Kontinuumstheorie der Versetzungen. Acta Mechanica 10, 59-66 (1970).

[399] M. Schechter: A boundary layer phenomenon in nonlinear membrane theory. SIAM Journal on Appl. Mathematics 14, 1099-1114 (1966).

[400] J. Schijeve: Brief note on couple-stresses. Natl. Luchten Ruimtevaartlab. Amsterdam, Rep. MP 226, 11p. (1964).

[401] J. Schijve: Note on couple-stresses. J. Mech. Phys. Solids 14, 113-119 (1966).

[402] J.A. Schouten: Ricci-Calculus (2nd edition). Berlin-Göttingen-Heidelberg, Springer (1954).

[403] L.I. Sedov: Über den Begriff des Spannungstensors bei Kontinuumsmodellen mit inneren Freiheitsgraden. ZAMP 20, 653-658 (1969).

[404] L.I. Sedov: Models of continuous media with internal degrees of freedom. PMM 32, 771-855 (1968). English translation in Appl. Math. Mech. 32, 803-818 (1969).

[405] D.E. Setzer: A plane elastostatic couple-stresses problem: extension of an elastic body containing a cut of finite length. Doct. Diss. Lehigh Univ. (1964).

[406] W.J. Shack: On linear viscoelastic rods. Int. J. Solids Structures 6, 1-20 (1970).

[407] M. Shimbot: An extension of the concept of Einsteinian coordinate with reference to asymmetric stress. RAAG Res. Notes (3rd ser.) 150, pp. 1-20 (1970).

[408] S.B. Sinha: Hypoelastic potential for bodies of grade two and one. Arch. Mech. Stos. 14, 171-180 (1962).

[409] J.C. Slattery: Spinning of a Noll simple fluid. A.I. Ch. E. Journal 12, 456-460 (1966).

[410] A.C. Smith: Deformations of micropolar elastic solids. Int. J. Engng. Sci. 5, 637-651 (1967).

[411] A.C. Smith: Waves in micropolar elastic solids. Int. J. Engng. Sci. 5, 741-746 (1967).

[412] A.C. Smith: Inequalities between the constants of a linear microelastic solid. Int. J. Engng. Sci. 6, 65-74 (1968).

[413] G.F. Smith: On rotational invariants of vectors and second-order tensors. Int. J. Engng. Sci. 8, 1-4 (1970).

[414] M. Sokolowski: Couple stresses in problems of torsion of prismatical bars. Bul. Acad. Pol. Sci.Sci. Techn. 13, 419-424 (1965).

[415] E. Soos: Uniqueness theorems for homogeneous, isotropic, simple elastic and thermoelastic materials having a microstructure. Int. J. Engng. Sci. 7, 257-268. (1969).

[416] E. Sternberg: Couple-stresses and singular stress concentrations in elastic solids. Mechanics of Generalized Continua. Proc. IUTAM Symposium Freudenstadt-Stuttgart 1967, pp. 95-108.

[417] E. Sternberg, R. Muki: The effect of couple-stresses on the stress-concentration around a crack. Int. J. Solids Structures 3, 69-95 (1967).

[418] R. Stojanović: Equilibrium conditions for internal stresses in non-Euclidean continua and stress spaces. Int. J. Engng. Sci. 1, 323-327 (1963).

[419] R. Stojanović: Dislocations in the generalized elastic Cosserat continuum. Mechanics of Generalized Continua. Proc. of the IUTAM Symposium Freudenstadt-Stuttgart 1967, pp. 152-155.

[420] R. Stojanović: On the stress relation in non-linear thermoelasticity. Non-linear Mechanics 4, 217-233 (1969).

[421] R. Stojanović: The elastic generalized Cosserat continuum with incompatible strains. Proc. of the Symp. on Fundamental Aspects of Dislocation Theory Washington, April 1969 pp. 817-830.

[422] R. Stojanović: A non-linear theory of thermoelasticity with couple-stresses. Trends in Elasticity and thermoelasticity. Nowacki Ann. Volume pp. 249-266. Groningen 1971.

[423] R. Stojanović, D. Blagojević: Couple-stresses. (In Serbocroatian). Dokumentacija. za gradjevinarstvo i arhitekturu, 883 (1966).

[424] R. Stojanović, D. Blagojević: On the general solution for torsion of polar elastic media. Int. J. Sol-

ids Structures 5, 251-260 (1969).

[425] R. Stojanović, S. Djurić: On the measures of strain in the theory of the elastic generalized Cosserat continuum. Symposia mathematica 1, 211-228 (1968).

[426] R. Stojanović, S. Djurić: Cosserat continua of grade two (In Serbocroatian) Proc. IX Yugoslav Congress of Mechanics, Split 1968, pp. 161-174.

[427] R. Stojanović, S. Djurić: The plane problem in the theory of the elastic oriented continua. (In Serbocroatian) Tehnika, 3, 388-393 (1969).

[428] R. Stojanović, S. Djurić, L. Vujosević: A contribution to the dynamics of Cosserat continua. (In Serbocroatian) Matematicki vesnik 1, (16) 127-140 (1964).

[429] R. Stojanović, S. Djurić, L. Vujosević: On Finite Thermal Deformations Arch. Mech. Stos. 16, 102-108 (1964).

[430] R. Stojanović, M. Gligorić: Incompatible strains of an elastic dielectric (In Serbocroatian) Proc. IX Yugoslav Congress of Mechanics Split 1968, pp. 175-186.

[431] R. Stojanović, L. Vujosević: Couple-stresses in non-Euclidean continua.Publ. Inst. Math. 2, (16), 71-74 (1962).

[432] R. Stojanović, L. Vujosević, D. Blagojević: A contribution to the theory of incompatible strains (In Serbocroatian).Proc. IX Yugoslav Congress of Mechanics, Split 1968, pp. 187-198.

[433] R. Stojanović, L. Vujosević, D. Blagojević: Couple-stress in thermoelasticity. Rev. Roum. Sci. Techn. Méc. Appl. 15, 517-537 (1970).

[434] R. Stojanovic, L. Vujosevic, S. Djuric: On the stress-strain relations for incompatible deformations. Plates and shells. Bratislava 1966. pp. 459-467.

[435] V.K. Stokes: Couple Stresses in Fluids. The Physics of Fluids 9, 1709-1715 (1966).

[436] J.A. Stuart, A.K. Kenneth: A theory of mixtures with microstructure. Zeitschrift für Angewandte Mathematik und Physik, Vol. 20, 1969 pp. 145-155.

[437] J. Sudria: L'action euclidienne de déformation et de mouvement. Mem. Sci. Physique 29, 56p., Paris (1935).

[438] E.S. Suhubi: Propagation of plane waves in an elastic medium with couple stresses. Bulteni Istanbul Teknik Universitesi 19, 138-152 (1966).

[439] E.S. Suhubi: Bending theory of microelastic bars. Bulteni Istanbul Teknik Universitesi 20, 19-28 (1967).

[440] E.S. Suhubi: On the foundation of the theory of rods. Int. J. Engng. Sci. 6, 169-191 (1968).

[441] E.S. Suhubi: Elastic dielectrics with polarization gradient. Int. J. Engng. Sci. 7, 993-997 (1969).

[442] E.S. Suhubi, C.A. Eringen: Nonlinear theory of microelastic solids - II Int. J. Engng. Sci. 2, 389-404 (1964).

[443] T.R. Tauchert, W.D. Clans Jr., T. Ariman: The linear theory of micropolar thermoelasticity. Int. J. Engng. Sci. 6, 37-47 (1968).

[444] P.P. Teodorescu: Sur l'action des charges concentrées dans le problème plan de la méchanique des solids déformables. Arch. Mech. Stosowanej 18, 567 (1966).

[445] P.P. Teodorescu: On the action of concentrated loads in the case of a Cosserat continuum. Mechanics of Generalized Continua. Proc. IUTAM Symposium Freudenstadt-Stuttgart 1967, pp. 120-125.

[446] P.P. Teodorescu, N. Sandru: Sur l'action des charges concentrées en elasticité asymmetrique plane. Rev. Roum. Math. Pures et Appl. 12, 1399-1405 (1967).

[447] P.P. Teodorescu: Sur la notion de moment massique dans le cas des corps du type de Cosserat. Bull. Acad. Pol. Sci.,Sci. Techn. 15, 57 (1967).

[448] P.P. Teodorescu: Sur les corps du type de Cosserat à l'élasticité linéaire. Symposia Mathematica 1, 375-409 (1968).

[449] G. Teodosiu: On the determination of internal stresses and couple-stresses in the continuum theory of dislocations. Bull. Acad. Sci. R.P.R. 12, 605 (1964).

[450] C. Teodosiu: The determination of stresses and couple-stresses generated by dislocations in isotropic media. Rev. Roum. Sci. Techn. Méc. Appl. 10, 1461-1480 (1965).

[451] C. Teodosiu: Continuous distribution of dislocations in hyperelastic materials of grade two. Mechanics of Generalized Continua. Proc. IUTAM Symposium Freudenstadt-Stuttgart 1967, pp. 279-822.

[452] C. Teodosiu: Contributions to the continuum theory of dislocations I,II Rev. Roum. Sci. Tech. Méc. Appl. 12, 961-977, 1061,1077 (1967).

[453] C. Teodosiu: Non-linear theory of materials of grade 2 with initial stresses and hyperstresses, I, II. Bul. Acad. Pol. Sci., Sci. Techn. 15, 95-102, 103-110 (1967).

[454] L.N. Ter-Mkrtich'ian: Homogeneous solution of two-dimensional problems of the theory of elasticity for a rectangular region of a Cosserat medium. Appl. Math. Mech. 33, 825-830 (1969) (translated from PMM 33, 850-854 (1969)).

[455] R. Tiffen, A.C. Stevenson: Elastic isotropy with body force and couple. Quart. J. Mech. Appl. Math. 9, 306-312 (1956).

[456] G.R. Tiwari: Effect of couple-stresses on the elastic stress distribution in a semi-infinite plate under a concentrated load. J. Sci. Engng. Res. India 9, 30-38 (1965).

[457] G.R. Tiwari: Effects of couple-stresses on the elastic stress distribution in an infinite plate having a circular hole. India J.Math. 8, 45-49 (1968).

[458] G.R. Tiwari: Effect of couple-stresses on the distribution of stresses along a crack. J. of Sci. and Engng. Res., India 12, 125-138 (1968).

[459] T. Tokuoka: Optical properties of polarizable linear micropolar fluids. Int. J. Engng. Sci. 8, 31-38 (1970).

[460] R.A. Toupin: The elastic dielectric. J. Rat. Mech. Anal. 5, 849-915 (1955).

[461] R.A. Toupin: Stress tensors in elastic dielectrics. Arch. Rat. Mech. Anal. 5, 1960, P 440-452.

[462] R.A. Toupin: Elastic materials with couple-stresses. Arch. Rat. Mech. Anal. 11, 385-414, (1962).

[463] R.A. Toupin: Theories of elasticity with couple-stress. Arch. Rat. Mech. Anal. 17, 85-112, (1964).

[464] R.A. Toupin: Dislocated and oriented media. Mechanics of Generalized Continua. Proc. IUTAM Symposium Freudenstadt-Stuttgart 1967, pp. 126-140.

[465] C. Truesdell: Principles of continuum mechanics. Colloquium Lectures on Pure and Appl. Sciences 5, Socony Mobil Oil Co. Dallas, Texas 1960.

[466] C. Truesdell: Thermodynamics for beginners. Proc. of IUTAM Symp. Vienna, 1966, pp. 373-389.

[467] C. Truesdell: The elements of continuum mechanics. Berlin-Heidelberg-New York, 1966.

[468] C. Truesdell, W. Noll: The nonlinear field theories of mechanics. Handbuch der Physik, Bd. III/3 (editor S. Flügge). Berlin-Heidelberg-New York, 1965.

[469] C. Truesdell, R. Toupin: The classical field theories. Handbuch der Physik Bd. III/1 (editor S. Flügge). Berlin-Göttingen-Heidelberg 1960.

[470] U. Ulhorn: Thermodynamics of a continuous system with internal structure. Proc. IUTAM Symposium Vienna 1966, pp. 390-393.

[471] A.J.C.B. de St. Venant: Mémoire sur la calcul de la résitance et de la flexion des piéces solides à simple ou à double courbure, en prenant simultanémént en considération les divers efforts auxquels elles peuvent être soumises dans tous les sens. C.R. Acad. Sci. Paris 17, 942-954, 1020-1031 (1843).

[472] L.P. Vinokurov, N.I.Derevyanke: Setting up of the basis equations for the analysis of rods (without torsion) taking into account couple-stresses (In Russian) PMM 2, 72-79 (1966).

[473] W. Voigt: Theoretische Studien über die Elastizitätsverhältnisse der Krystalle. Abh. Wiss. Ges. Göttingen, 34 (1887).

[474] W. Voigt: Kompendium der theoretischen Physik, Bd. 1, Leipzig 1895/6.

[475] L. Vujošević: Internal stresses of a continuum with incompatible strains. (In Serbocroatian) Proc. IX Yugoslav Congress of Mechanics, Split 1968, pp. 71-84.

[476] L. Vujošević: On the elasticity coefficients for some materials with incompatible strains. (In Serbocroatian) Proc. IX Yugoslav Congress of Mechanics, Split 1968, pp. 71-84.

[477] L. Vujošević, R. Stojanovic: Linear incompatible deformations of materials with cubic symmetry. (In Serbocroatian) Tehnika 7, 117-121 (1969).

[478] L. Vujošević, R. Stojanovic: A second-order theory of incomaptible elastic deformations. (In Serbocroatian) To appear in Tehnika, 1969.

[479] F.Y.M. Wan: Stress functions for nonlinear shallow shell theory. Studies in Appl. Math. 48, 177-179 (1969).

[480] F.Y.M. Wan: Exact reductions of the equations of linear theory of shells of revolution. Studies in Appl. Math. 48, 361-375 (1969).

[481] F.Y.M. Wan: On the equations of the linear theory of elastic conical shells. Studies in Appl. Math. 49, 69-83 (1970).

[482] C.C. Wang: A general theory of subfluids. Arch. Rat. Mech. Anal. 20, 1 (1965).

[483] C.C. Wang: On the geometric structure of simple bodies, a mathematical foundation for the theory of continuous distribution of dislocations. Arch. Rat. Mech. Anal. 27, 33-94 (1967).

[484] C.C. Wang: Generalized simple bodies. Arch. Rat. Mech. Anal. 32, 1-29 (1969).

[485] Y. Weitsman: Couple-stress effects on stress concentration around a cylindrical inclusion in a field of uniaxial tension. J. Appl. Mech. (Trans. ASME) E 32, 424-428 (1965).

[486] Y. Weitsman: Strain gradient effects around cylindrical inclusions and cavities in a field of cylindrically symmetric tension. J. Appl. Mech. Trans. ASME, March 1966, pp. 57-67.

[487] Y. Weitsman: A note on singularities in a Cosserat continuum. Quart. Appl. Math. 25, 213-217 (1967).

[488] Y. Weitsman: Initial stresses and skin effects in a hemitropic Cosserat continuum. J. Appl. Mech. (Trans. ASME) E 34, 160-164 (1967).

[489] Y. Weitsman: Two dimensional singular solutions in infinite regions with couple-stresses. Quart. Appl. Math. 25, 485-489 (1968).

[490] G. Wempner: New concepts for infinite elements of shells. ZAMM 48, T 174 (1968).

[491] M.L. Wenner: On torsion of an elastic cylindrical Cosserat surface. Int. J. Solids Structures 4, 769-776 (1968).

[492] Z. Wesolowski: On the couple-stresses in an elastic continuum. Arch. Mech. Stosow. 17, 219-232 (1965).

[493] A.B. Whitman, C.N. DeSilva: A dynamical theory of elastic directed curves. ZAMP 20, 200-213 (1969).

[494] A.B. Whitman, C.N. DeSilva: Dynamics and stability of elastic Cosserat curves. Int. J. Solids Structures 6, 411-422 (1970).

[495] K. Wieghardt: Statische und dynamische Versuche mit Sand als Strommungsmedium ZAMM 48, 7247 (1968).

[496] A.J. Willson: The flow of a micropolar liquid layer down an inclined plane. Proc. Camb. Phil. Soc. 64, 513-526 (1968).

[497] K. Wilmanski: Asymptotic methods in the theory of a body with microstructure. The plane problem. Rozprawy Inzynierskie, 15, 295-309 (1967).

[498] K. Wilmanski, C. Wozniak: On geometry of continuous medium with micro-structure. Arch. Mech. Stosow. 19, 715 (1967).

[499] C. Wozniak: Theory of fibrous media I, II Arch. Mech. Stosow. 17, 651-670, 779-799 (1965).

[500] C. Wozniak: Foundation of Micro-solid Theory. Bull. Acad. Pol. Sci., Sci. Techn. 14, 567 (1966).

[501] C. Wozniak: Thermoelasticity of Micro-materials. Bull. Acad. Pol. Sci., Sci. Techn. 14, 573 (1966).

[502] C. Wozniak: Hyperstresses in linear Thermoelasticity. Bull. Acad. Pol. Sci., Sci. Techn., 14, 637-642 (1966).

[503] C. Wozniak: State of stress in the lattice-type bodies. Bull. Acad. Pol. Sci., Sci. Techn., 14, 643 (1966

[504] C. Wozniak: Thermoelasticity of non-simple oriented materials. Int. J. Engng. Sci. 5, 605-612 (1967).

[505] C. Wozniak: Foundations of dynamics of continua I, II. Bul. Acad. Pol. Sci., Sci. Techn. 15, 435-442, 443-446 (1967).

[506] C. Wozniak: Thermoelasticity of bodies with microstructures. Arch. Mech. Stosowanej 19, 335-365 (1967).

[507] C. Wozniak: Theory of thermoelasticity of non-simple materials. Arch. Mech. Stosowanej 19, 485-493 (1967).

[508] C. Wozniak: Introduction to dynamics of deformable bodies. Arch. Mech. Stosow. 19, 627-664 (1967).

[509] C. Wozniak: Conduction of heat in lattice type structures. Arch. Mech. Stosow. 21, 495-505 (1969).

[510] C. Wozniak: On the equations of the theory of lattice structures. Arch. Mech. Stosow. 21, 539-555 (1969).

[511] J. Wyrwinski: Green functions for a thermoelastic Cosserat medium. Bull. Acad. Pol. Sci., Sci. Techn. 14, 113-122 (1966).

[512] Y. Yamamoto: An intrinsic theory of Cosserat continuum. Int. J. Solids Structures 4, 1013-1024 (1968).

[513] S. Zahorski: Plane sound waves in non-simple elastic materials. Bull. Acad. Pol. Sci., Sci. Techn. 15, 339-343 (1967).

[514] S. Zahorski: Plane sound waves in non-simple media of arbitrary grade. Arch. Mech. Stosow. 19, 551-566 (1967).

[515] V.A. Zhelnorović: Modeli sred s vnutreniim electromagnitnjim i mehaniceskin momentami. Problem'i gidrodinamiki i mehaniki splosnoi sred'i. Sedov Anniversary Volume, Moskva 1969. Izdateljstvo "Nauka". pp. 221-231.

[516] H. Ziegler: Some extremeum principles in irreversible thermodynamics with applications to continuum mechanics. Progress in Solid Mechanics IV (editors I. Sneddon and R. Hill) pp. 93-198. Amsterdam 1963.

[517] H. Zorski: On the structure of the stress tensor in the continuous medium. (In Russian) 2-j Vses. Sezd. po Tero. i Prikl. Mekh. 1964, pp. 243-244.

Contents

Printed in the United States
by Bookmasters

Printed in the United States
By Bookmasters